Lecture Notes in Mathematics

Edited by J.-M. Morel, F. Takens and B. Teissier

Editorial Policy
for the publication of monographs

1. Lecture Notes aim to report new developments in all areas of mathematics and their applications – quickly, informally and at a high level. Mathematical texts analysing new developments in modelling and numerical simulation are welcome.

   Monograph manuscripts should be reasonably self-contained and rounded off. Thus they may, and often will, present not only results of the author but also related work by other people. They may be based on specialised lecture courses. Furthermore, the manuscripts should provide sufficient motivation, examples and applications. This clearly distinguishes Lecture Notes from journal articles or technical reports which normally are very concise. Articles intended for a journal but too long to be accepted by most journals, usually do not have this „lecture notes" character. For similar reasons it is unusual for doctoral theses to be accepted for the Lecture Notes series, though habilitation theses may be appropriate.

2. Manuscripts should be submitted (preferably in duplicate) either to Springer's mathematics editorial in Heidelberg, or to one of the series editors (with a copy to Springer). In general, manuscripts will be sent out to 2 external referees for evaluation. If a decision cannot yet be reached on the basis of the first 2 reports, further referees may be contacted: The author will be informed of this. A final decision to publish can be made only on the basis of the complete manuscript, however a refereeing process leading to a preliminary decision can be based on a pre-final or incomplete manuscript. The strict minimum amount of material that will be considered should include a detailed outline describing the planned contents of each chapter, a bibliography and several sample chapters.

   Authors should be aware that incomplete or insufficiently close to final manuscripts almost always result in longer refereeing times and nevertheless unclear referees' recommendations, making further refereeing of a final draft necessary.

   Authors should also be aware that parallel submission of their manuscript to another publisher while under consideration for LNM will in general lead to immediate rejection.

3. Manuscripts should in general be submitted in English. Final manuscripts should contain at least 100 pages of mathematical text and should always include

   – a table of contents;

   – an informative introduction, with adequate motivation and perhaps some historical remarks: it should be accessible to a reader not intimately familiar with the topic treated;

   – a subject index: as a rule this is genuinely helpful for the reader.

   For evaluation purposes, manuscripts may be submitted in print or electronic form (print form is still preferred by most referees), in the latter case preferably as pdf- or zipped ps-files. Lecture Notes volumes are, as a rule, printed digitally from the authors' files. To ensure best results, authors are asked to use the LaTeX2e style files available from Springer's web-server at:

   ftp://ftp.springer.de/pub/tex/latex/mathegl/mono/ (for monographs) and

   ftp://ftp.springer.de/pub/tex/latex/mathegl/mult/ (for summer schools/tutorials).

   Additional technical instructions, if necessary, are available on request from lnm@springer-sbm.com.

Lecture Notes in Mathematics

1921

Editors:
J.-M. Morel, Cachan
F. Takens, Groningen
B. Teissier, Paris

Jianjun Paul Tian

Evolution Algebras
and their Applications

 Springer

Author

Jianjun Paul Tian

Mathematical Biosciences Institute
The Ohio State University
231 West 18th Avenue
Columbus, OH 43210-1292
USA

Contact after August 2007

Mathematics Department
College of William and Mary
P. O. Box 8795
Williamsburg VA 23187-8795
USA

e-mail: jtian@wm.edu

Library of Congress Control Number: 2007933498

Mathematics Subject Classification (2000): 08C92, 17D92, 60J10, 92B05, 05C62, 16G99

ISSN print edition: 0075-8434
ISSN electronic edition: 1617-9692
ISBN 978-3-540-74283-8 Springer Berlin Heidelberg New York
DOI 10.1007/978-3-540-74284-5

Springer is a part of Springer Science+Business Media
springer.com
© Springer-Verlag Berlin Heidelberg 2008

Typesetting by the author and SPi using a Springer LATEX macro package

Cover design: *design & production* GmbH, Heidelberg

Printed on acid-free paper   SPIN: 12109265   41/SPi   5 4 3 2 1 0

To my parents

Bi-Yuan Tian and Yu-Mei Liu

My father, the only person I know who can operate two abaci
using his left and right hand simultaneously in his business.

Preface

In this book, we introduce a new type of algebra, which we call evolution algebras. These are algebras in which the multiplication tables are of a special type. They are motivated by evolution laws of genetics. We view alleles (or organelles or cells, etc,) as generators of algebras. Therefore we define the multiplication of two "alleles" G_i and G_j by $G_i \cdot G_j = 0$ if $i \neq j$. However, $G_i \cdot G_i$ is viewed as "self-reproduction," so that $G_i \cdot G_i = \sum_j p_{ij} G_j$, where the summation is taken over all generators G_j. Thus, reproduction in genetics is represented by multiplication in algebra. It seems obvious that this type of algebra is nonassociative, but commutative. When the p_{ij}s form Markovian transition probabilities, the properties of algebras are associated with properties of Markov chains. Markov chains allow us to develop an algebra theory at deeper hierarchical levels than standard algebras. After we introduce several new algebraic concepts, particularly algebraic persistency, algebraic transiency, algebraic periodicity, and their relative versions, we establish hierarchical structures for evolution algebras in Chapter 3. The analysis developed in this book, particularly in Chapter 4, enables us to take a new perspective on Markov process theory and to derive new algebraic properties for Markov chains at the same time. We see that any Markov chain has a dynamical hierarchy and a probabilistic flow that is moving with invariance through this hierarchy. We also see that Markov chains can be classified by the skeleton-shape classification of their evolution algebras. Remarkably, when applied to non-Mendelian genetics, particularly organelle heredity, evolution algebras can explain establishment of homoplasmy from heteroplasmic cell population and the coexistence of mitochondrial triplasmy, and can also predict all possible mechanisms to establish the homoplasmy of cell population. Actually, these mechanisms are hypothetical mechanisms in current mitochondrial disease research. By using evolution algebras, it is easy to identify different genetic patterns from the complexity of the progenies of *Phytophthora infectans* that cause the late blight of potatoes and tomatoes. Evolution algebras have many connections with other fields of mathematics, such as graph theory, group theory, knot theory, 3-manifolds, and Ihara-Selberg zeta functions. Evolution algebras provide a theoretical framework to unify many phenomena. Among the further research topics related to evolution algebras and other fields, the most significant topic perhaps is to develop a continuous evolution algebra theory for continuous time dynamical systems.

The intended audience of this book includes graduate students and researchers with interest in theoretical biology, genetics, Markov processes, graph theory, and nonassociative algebras and their applications.

Professor Jean-Michel Morel gave me a lot of support and encouragement, which enabled me to take the step to publish my research results as a book. Other editors and staff in LNM made efforts to find reviewers and edit my book. Here, I wish to express my great thanks to them.

I thank Professor Michael T. Clegg for his stimulating problems in coalescent theory. From that point, I began to study genetics and stochastic processes. I am greatly indebted to Professor Xiao-Song Lin, my Ph.D advisor, for his valuable advice and long-time guidance. I am thankful to professors Bai-Lian Larry Li, Michel L. Lapidus, and Barry Arnold for their valuable suggestions. It gives me great pleasure to thank Professors Bun Wong, Yat Sun Poon, Shizhong Xu, Keh-Shin Lii, Peter March, Dennis Pearl, Raymond L. Orbach, Murray Bremner, Yuan Lou, and Yang Kuang for their encouragement. I also thank Professor C. William Birky Jr. for his explanation of non-Mendelian genetics through e-mails. I acknowledge Professor Winfried Just for his suggestions of writing style of the book and a formula in Chapter 3. I am grateful to my current mentor, Professor Avner Friedman, for his detailed and cherished suggestions on the research in this book and my other research directions. I thank three reviewers for their suggestions and constructive comments.

Last, but not the least, I thank Dr. Shannon L. LaDeau for her help on English of the book. I also thank my wife, Yanjun Sophia Li, for her support and love. I acknowledge the support from the National Science Foundation upon agreement No. 0112050.

Mathematical Biosciences Institute, Ohio   *Jianjun Paul Tian*
                                                            April, 2007

Contents

1

Introduction

While I was studying stochastic processes and genetics, it occurred to me that there exists an intrinsic and general mathematical structure behind the neutral Wright-Fisher models in population genetics, the reproduction of bacteria involved by bacteriophages, asexual reproduction or generally non-Mendelian inheritance, and Markov chains. Therefore, we defined it as a type of new algebra — the evolution algebra. Evolution algebras are nonassociative and non-power-associative Banach algebras. Indeed, they are natural examples of nonassociative complete normed algebras arising from science. It turns out that these algebras have many unique properties, and also have connections with other fields of mathematics, including graph theory (particularly, random graphs and networks), group theory, Markov processes, dynamical systems, knot theory, 3−manifolds, and the study of the Riemann-zeta function (or a version of it called the Ihara-Selberg zeta function). One of the unusual features of evolution algebras is that they possess an evolution operator. This evolution operator reveals the dynamical information of evolution algebras. However, what makes the theory of evolution algebras different from the classical theory of algebras is that in evolution algebras, we can have two different types of generators: algebraically persistent generators and algebraically transient generators.

The basic notions of algebraic persistency and algebraic transiency, and their relative versions, lead to a hierarchical structure on an evolution algebra. Dynamically, this hierarchical structure displays the direction of the flow induced by the evolution operator. Algebraically, this hierarchical structure is given in the form of a sequence of semidirect-sum decompositions of a general evolution algebra. Thus, this hierarchical structure demonstrates that an evolution algebra is a mixed algebraic and dynamical subject. The algebraic nature of this hierarchical structure allows us to have a rough skeleton-shape classification of evolution algebras. At the same time, the dynamical nature of this hierarchical structure is what makes the notion of evolution algebra applicable to the study of stochastic processes and many other subjects in different fields. For example, when we apply the structure theorem to the evolution algebra induced by a Markov chain, it is easy to see that the Markov chain has a dynamical hierarchy and the probabilistic flow is moving with invariance through this hierarchy, and that all Markov chains can be classified by the skeleton-shape classification of their induced evolution algebras. Hierarchical structures of Markov chains may be stated in other terms. But, it is the first time that we show algebraic properties of Markov chains and a complete skeleton-shape classification of Markov chains. Although evolution algebra theory is an abstract system, it gives insight into the understanding of non-Mendelian genetics. For instance, once we apply evolution algebra theory to the inheritance of organelle genes, we can predict all possible mechanisms to establish the homoplasmy of cell populations. Actually, these mechanisms are hypothetical mechanisms in current mitochondrial research. Using our algebra theory, it is also easy to understand the coexistence of triplasmy in tissues of sporadic mitochondrial disorder patients. Further more, once the algebraic structure of asexual progenies of *Phytophthora infectans* is obtained, we can make certain important predictions and suggestions to plant pathologists.

In history, mathematicians and geneticists once used nonassociative algebras to study Mendelian genetics. Mendel [30] first exploited symbols that are quite algebraically suggestive to express his genetic laws. In fact, it was later termed "Mendelian algebras" by several other authors. In the 1920s and 1930s, general genetic algebras were introduced. Apparently, Serebrowsky [31] was the first to give an algebraic interpretation of the sign "×", which indicated sexual reproduction, and to give a mathematical formulation of Mendel's laws. Glivenkov [32] introduced the so-called Mendelian algebras for diploid populations with one locus or two unlinked loci. Independently, Kostitzin [33] also introduced a "symbolic multiplication" to express Mendel's laws. The systematic study of algebras occurring in genetics can be attributed to I. M. H. Etherington. In his series of papers [34], he succeeded in giving a precise mathematical formulation of Mendel's laws in terms of nonassociative algebras. Besides Etherington, fundamental contributions have been made by Gonshor [35], Schafer [36], Holgate [37, 38], Hench [39], Reiser [40], Abraham [41], Lyubich [47], and Worz-Busekos [46]. It is worth mentioning two unpublished work in the field. One is the Ph.D. thesis of Claude Shannon, the founder of modern information theory, which was submitted in 1940 (The Massachusetts Institute of Technology) [43]. Shannon developed an algebraic method to predict the genetic makeup in future generations of a population starting with arbitrary frequencies. The other one is Charles Cotterman's Ph.D. thesis that was also submitted in 1940 (The Ohio State University) [44] [45]. Cotterman developed a similar system as Shannon did. He also put forward a concept of derivative genes, now called "identical by descent."

During the early days in this area, it appeared that the general genetic algebras or broadly defined genetic algebras, could be developed into a field of independent mathematical interest, because these algebras are in general not associative and do not belong to any of the well-known classes of nonassociative algebras such as Lie algebras, alternative algebras, or Jordan algebras.

They possess some distinguishing properties that lead to many interesting mathematical results. For example, baric algebras, which have nontrivial representations over the underlying field, and train algebras, whose coefficients of rank equations only are functions of the images under these representations, are new concepts for mathematicians. Until 1980s, the most comprehensive reference in this area was Worz-Busekros's book [46]. More recent results, such as genetic evolution in genetic algebras, can be found in Lyubich's book [47]. A good survey is Reed's article [48].

General genetic algebras are the product of interaction between biology and mathematics. Mendelian genetics introduced a new subject to mathematics: general genetic algebras. The study of these algebras reveals algebraic structures of Mendelian genetics, which always simplifies and shortens the way to understand genetic and evolutionary phenomena. Indeed, it is the interplay between purely mathematical structures and the corresponding genetic properties that makes this area so fascinating. However, after Baur [49] and Correns [50] first detected that chloroplast inheritance departed from Mendel's rules, and much later, mitochondrial gene inheritance was also identified in the same way, and non-Mendelian inheritance of organelle genes was recognized with two features — uniparental inheritance and vegetative segregation. Now, non-Mendelian genetics is a basic language of molecular geneticists. Logically, we can ask what non-Mendelian genetics offers to mathematics. The answer is "evolution algebras" [24].

The purpose of the present book is to establish the foundation of the framework of evolution algebra theory and to discuss some applications of evolution algebras in stochastic processes and genetics. Obviously, we are just opening a door to a new subject of the mixture of algebras and dynamics and to the many new research topics that are confronting us. To promote further research in this subject, we include many specific research topics and open problems at the end of this book. Now, I would like to briefly introduce the content contained in each chapter of the book.

In Chapter 2, we introduce the motivations behind the study of evolution algebras from the perspective of three different sciences: biology, physics, and mathematics. We observe phenomena of uniparental inheritance and the reproduction of bacteria involved by bacteriophages; we also analyze the neutral Wright-Fisher model for a haploid population in population genetics. We study motions of particles in a space and discrete flows in a discrete space, and we also observe reactions among particles in general physics. We mention some research in knot theory where negative probabilities are involved. We analyze and view a Markov chain as a discrete time dynamical system. All these phenomena suggest a common and intrinsic algebraic structure, which we define in chapter 3 as evolution algebras.

In Chapter 3, evolution algebras are defined; their basic properties are investigated and the principal theorem about evolution algebras — the hierarchical structure theorem — is established. We define evolution algebras in terms of generators and defining relations. Because the defining relations are unique for an evolution algebra, the generator set can serve as a basis for an evolution algebra. This property gives some advantage in studying evolution algebras. The basic algebraic properties of evolution algebras, such as nonassociativity and nonpower-associativity are studied. Various algebraic concepts in evolution algebras are also investigated, such as evolution subalgebras, the associative multiplication algebra of an evolution algebra, the centroid of an evolution algebra and, the derived Lie algebra of an evolution algebra. The occurrence relation among generators of an evolution algebra and the connectedness of an evolution algebra are defined. We utilize the occurrence relation to define the periodicity of generators. From the viewpoint of dynamical systems, we introduce an evolution operator for an evolution algebra that is actually a special right (left) multiplication operator. This evolution operator reveals the dynamical information of an evolution algebra. To describe the evolution flow quantitatively, we introduce a norm for an evolution algebra. Under this norm, an evolution algebra becomes a Banach algebra. As we have mentioned above, what makes the evolution algebra theory different from the classical algebra theory is that in evolution algebras we can have two different categories of generators, algebraically persistent generators and algebraically transient generators. Moreover, the difference between algebraic persistency and algebraic transiency suggests a direction of dynamical flow as it displays in the hierarchy of an evolution algebra. The remarkable property of an evolution algebra is its hierarchical structure, which gives a picture of a dynamical process when one takes multiplication in an evolution algebra as time-step in a discrete-time dynamical system. Algebraically, this hierarchy is a sequence of semidirect-sum decompositions of a general evolution algebra. It depends upon the "relative" concepts of algebraic persistency and algebraic transiency. By "relative" concepts, we mean that concepts of higher level algebraic persistency and algebraic transiency are defined over the space generated by transient generators in the previous level. The difference between algebraic persistency and algebraic transiency suggests a sequence of the semidirect-sum decompositions, or suggests a direction of the evolution from the viewpoint of dynamical systems. This hierarchical structure demonstrates that an evolution algebra is a mixed subject of algebras and dynamics. We also obtain the structure theorem for a simple evolution algebra. We give a way to reduce a "big" evolution algebra to a "small" one that still has the same hierarchy as that of the original algebra. We call it the reducibility. This reducibility gives a rough classification, the skeleton-shape classification, of all evolution algebras.

To demonstrate the importance and the applicability of the abstract subject — evolution algebras — we study a type of evolution algebra that corresponds to or is determined by a Markov chain in Chapter 4. We see that any general Markov chain has a dynamical hierarchy and the probabilistic flow is moving with invariance through this hierarchy, and that all Markov chains can be classified by the skeleton-shape classification of their evolution algebras. When a Markov chain is viewed as a dynamical system, there should be a certain mechanism behind the Markov chain. We view this mechanism as a "reproduction process." But it is a very special case of reproduction process. Each state can just "cross" with itself, and different states cannot cross, or they cross to produce nothing. We introduce a multiplication for this reproduction process. Thus an evolution algebra is defined by using transition probabilities of a Markov chain as structural constants. In evolution algebras, the Chapman-Kolmogorov equations can be simply viewed as a composition of evolution operators or the principal power of a special element. By using evolution algebras, one can see algebraic properties of Markov chains. For example, a Markov chain is irreducible if and only if its evolution algebra is simple, and a subset of state space of a Markov chain is closed in the sense of probability if and only if it generates an evolution subalgebra. An element has the algebraic period of d if and only if it has the probabilistic period of d. Generally, a generator is probabilistically transient if it is algebraically transient, and a generator is algebraically persistent if it is probabilistically persistent. When the dimension of the evolution algebra determined by a Markov chain is finite, algebraic concepts (algebraic persistency and algebraic transiency) and analytic concepts (probabilistic persistency and probabilistic transiency) are equivalent. We also study the spectrum theory of the evolution algebra M_X determined by a Markov chain X. Although the dynamical behavior of an evolution algebra is embodied by various powers of its elements, the evolution operator seems to represent a "total" principal power. From the algebraic viewpoint, we study the spectrum of evolution operators. Particularly, the evolution operator is studied at the $0th$ level in the hierarchy of an evolution algebra. For example, for a finite dimension evolution algebra the geometric multiplicity of the eigenvalue 1 of the evolution operator is equal to the number of the $0th$ simple evolution subalgebras. The spectrum structure at higher level is an interesting further research topic. Another possible spectrum theory could be the study of plenary powers. Actually, we have already defined the plenary power for a matrix. It could give a way to study this possible spectrum theory. Any general Markov chain has a dynamical hierarchy, which can be obtained from its corresponding evolution algebra. We give a description of probability flows on its hierarchy. We also give the sojourn times during each simple evolution subalgebra at each level on the hierarchy. By using the skeleton-shape classification of evolution algebras, we can reduce a bigger Markov chain to a smaller one that still possesses the same dynamical behavior as the original chain does. We have also obtained a new skeleton-shape classification theorem for general Markov chains. Thus, from the evolution algebra theory, algebraic properties about general Markov chains are revealed. In the last section of this chapter, we discuss examples and applications, and show algebraic versions of Markov chains, evolution algebras, also have advantages in computation of Markov processes.

We begin to apply evolution algebra theory to biology in Chapter 5. We first introduce the basic biology of non-Mendelian genetics including organelle population genetics and *Phytophthora infectans* population genetics.

We then give a general algebraic formulation of non-Mendelian inheritance. To understand a puzzling feature of organelle heredity, that is that heteroplasmic cells eventually disappear and the homoplasmic progenies are observed, we construct relevant evolution algebras. We then can predict all possible mechanisms to establish the homoplasmy of cell populations, which actually are hypothetical mechanisms in current mitochondrial research [55]. Theoretically, we can discuss any number of mitochondrial mutations and study their genetic dynamics by using evolution algebras. Remarkably, experimental biologists have observed the coexistence of the triplasmy (partial duplication of mt-DNAs, deletion of mt-DNAs, and wild-type mt-DNAs) in tissues of patients with sporadic mitochondrial disorders. While doctors and biologists cultured cell lines to study the dynamical relations among these mutants of mitochondria, our algebra model could be used to predict the outcomes of their cell line cultures. We show that concepts of algebraic transiency and algebraic persistency catch the essences of biological transitory and biological stability. Moreover, we could predict some transition phases of mutations that are difficult to observe in experiments. We also study another type of uniparental inheritance about *Phytophthora infectans* that cause late blight of potatoes and tomatoes. After constructing several relevant evolution algebras for the progeny populations of *Phytophthora infectans*, we can see different genetically dynamical patterns from the complexity of the progenies of *Phytophthora infectans*. We then predict the existence of intermediate transient races and the periodicity of reproduction of biological stable races. Practically, we can help farmers to prevent spread of late blight disease. Theoretically, we can use evolution algebras to provide information on *Phytophthora infectans* reproduction rates for plant pathologists.

As we mentioned above, evolution algebras have many connections with other fields of mathematics. Using evolution algebras it is expected that we will be able to see problems in many mathematical fields from a new perspective. We have already finished some of the basic study. Most of the research will be very interesting and promising both in theory and in application. To promote better understanding and further research in evolution algebras, in Chapter 6, we list some of the related results we have obtained and put forward further research topics and open problems. For example, we obtain a theorem of classification of directed graphs. We also post a series of open problems about evolution algebras and graph theory. Because evolution algebras hold the intrinsic and coherent relation with graph theory, we will be able to analyze graphs algebraically. The purpose of this is that we try to establish a brand new theory "algebraic graph theory" to reach the goal of Gian-Carlo Rota — "Combinatorics needs fewer theorems and more theory" [29]. On the other hand, it is also expected that graph theory can be used as a tool to study nonassociative algebras. Some research topics in evolution algebras and group theory, knot theory, and Ihara-Selberg zeta function, which we post as further research topics, are also very interesting. Perhaps, the most significant topic is to develop a continuous evolution algebra theory for continuous time dynamical systems. It is also important to use evolution algebras to develop algebraic statistical physics models. In this direction, the big picture in our mind is to describe the general interaction of particles. This means any two generators can multiply and do not vanish when they are different. This involves an operation, multiplication, of three-dimensional matrices. Some preliminary results have already been obtained in this direction. We are also interested in questions such as how evolution algebras reflect properties of a 3-manifold where a particle moves when the recording time period is taken as an infinite sequence, and what new results about the 3-manifold can be obtained by the sequence of evolution algebras, etc.

We give a list of background literature in the last section, though the directly related literature is sparse.

2

Motivations

In this chapter, we provide several examples from biology, physics, and mathematics including topology and stochastic processes, which have motivated the development of the theory of evolution algebras.

2.1 Examples from Biology

2.1.1 Asexual propagation

Prokaryotes are nonsexual reproductive organisms. Prokaryotic cells, unlike eukaryotic cells, do not have nuclei. The genetic material (DNA) is concentrated in a region called the nucleoid, with no membrane to separate this region from the rest of the cell. In prokaryote inheritance, there is no mitosis and meiosis. Instead, prokaryotes reproduce by binary fission. That is, after the prokaryotic chromosome duplicates and the cell enlarges, the enlarged cell becomes two small cells divided by a cell wall. Basically, the genetic information passed from one generation to the next should be conserved because of the strictness of DNA self-replication. However, there are still many possible factors in the environment that can induce the change of genetic information from generation to generation. The inheritance of prokaryotes is then not Mendelian. The first factor is DNA mutation. The second factor is related to gene recombination between a prokaryotic gene and a viral gene, for example bacteriophage λ's gene. This process of recombination between a prokaryotic gene and a viral gene is called gene transduction. For the detailed process of transduction, please refer to Nell Campbell [15]. The third factor comes from conjugation induced by sex plasmids. That is a direct transfer of genetic material between two prokaryotic cells. The most extensively studied case is Escherichia coli. Figure 2.1 depicts the division of bacterial cell from the book [15].

Now, let's mathematically formulate the asexual reproduction process. Suppose that we have n genetically distinct prokaryotes, denoting them by

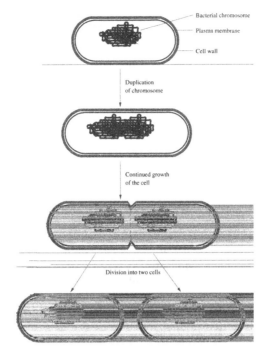

Fig. 2.1. Bacterial cell division

$p_1, p_2, \ldots, p_n$. We also suppose that the same environmental conditions are maintained from generation to generation. We look at changes in gene frequencies over two generation. We can view it either from the population standpoint or from the individual standpoint. To this end, we can set the following relations:

$$\begin{cases} p_i \cdot p_i = \sum_{k=1}^{n} c_{ik} p_k, \\ \quad p_i \cdot p_j = 0, i \neq j. \end{cases}$$

Here, we view the multiplication as asexual reproduction.

2.1.2 Gametic algebras in asexual inheritance

Let us recall some basic facts in general genetic algebras first [22]. Consider an infinitely large, randomly mating population of diploid individuals, with individuals differing genetically at one or several autosomal loci. Let $a_1, a_2, \ldots, a_n$ be the genetically distinct gametes produced by the population. By random union of gametes a_i and a_j, zygotes of type $a_i a_j$ are formed. Assume that a zygote $a_i a_j$ produces a number γ_{ijk} of gametes of type a_k, which survive in the next generation, $k, i, j = 1, 2, \ldots, n$. In the absence of selection, we assume all zygotes have the same fertility, and every zygote produces the same number of surviving gametes. Thus, one can have the probability that a zygote $a_i a_j$

produces a gamete a_k by number γ_{ijk}, still denoting γ_{ijk} as the probability that satisfies $\sum_{k=1}^n \gamma_{ijk} = 1$. The frequency of gamete a_k produced by the total population is $\sum_{i,j=1}^n v_i \gamma_{ijk} v_j$ if the gamete frequency vector of parental generation is $(v_1, v_2, \ldots, v_n)$. Now, the gamete algebra is defined on the linear space spanned by these gametes $a_1, a_2, \ldots, a_n$ over the real number field by the following multiplication table

$$a_i a_j = \sum_{k=1}^n \gamma_{ijk} a_k, \quad i, j = 1, 2, \ldots, n,$$

and then linear extension onto the whole space. However, when we consider the asexual inheritance, the interpretation $a_i a_j$ as a zygote does not make sense biologically if $a_i \neq a_j$. But, $a_i a_i = a_i^2$ can still be interpreted as self-replication. Therefore, in asexual inheritance, we can use the following relations to define an algebra

$$\begin{cases} a_i \cdot a_i = \sum_{k=1}^n \gamma_{ik} a_k, \\ a_i \cdot a_j = 0, \quad i \neq j. \end{cases}$$

In the asexual inheritance, $a_i a_j$ is no longer a zygote; actually, it does not exist. Mathematically, we set $a_i a_j = 0$. Of course, this case is not of Mendelian inheritance.

2.1.3 The Wright-Fisher model

In population genetics, one often considers evolutionary behavior of a diploid population with a fixed size N. Suppose that the individuals in this population are monoecious and that no selective differences exist between two alleles A_1 and A_2 possible at a certain locus A. There are, $g_1, g_2, \ldots, g_n$, $n = 2N$ genes in the population in any generation. If we do not pay attention to genealogical relations, it is sufficient to know the number X of A_1 gene in each generation for understanding population evolutionary behavior. Clearly in any generation, X takes one of the values $0, 1, \ldots, 2N$, and we denote the value assumed by X in generation t by $X(t)$. We must assume some specific model that describes the way in which the genes in generation $t+1$ are derived from the genes in generation t. The Wright-Fisher model [2] [16] assumes that the genes in generation $t + 1$ are derived by sampling with replacement from the genes of generation t. This means that the number $X(t + 1)$ is a binomial random variable with index n and parameter $\frac{X(t)}{n}$. More explicitly, given $X(t) = k$, the probability p_{kl} of $X(t + 1) = l$ is given by

$$p_{kl} = \binom{n}{l} \left(\frac{k}{n}\right)^l \left(1 - \frac{k}{n}\right)^{n-l}.$$

It is clear that $X(t)$ has markovian properties. Now, if we just overlook the details of the reproduction process and consider these probabilities as numbers, we may say that a certain gene, name it g_i in generation t, can reproduce

p_{ij} genes g_j in generation $t + 1$. So, we focus on each individual gene to study its reproduction from the population level. Of course, the crossing of genes does not make any sense genetically, although the "replication" of a gene has certain biological meanings. Therefore, this viewpoint suggests the following symbolical formulae

$$\begin{cases} g_i \cdot g_i = \sum_{j=1}^n m_{ij} g_j \\ g_i \cdot g_j = 0, \quad i \neq j \end{cases},$$

where m_{ij} is the number of "offspring" of g_i. We will study a simple case that includes selection as a parameter in Example 7.

2.2 Examples from Physics

2.2.1 Particles moving in a discrete space

Consider a particle moving in a discrete space, for example, in a graph G. Suppose it starts at vertex v_i, then, which vertex will be its second position depends on which neighbor of v_i this particle prefers to. We may attach a preference coefficient to each edge from v_i to its neighbor v_j. For instance, we use w_{ij} as the preference coefficient, which is not necessarily a probability. Thus, the second position will be the vertex that this particle most prefers to. This particle will move on the graph continuously. If the particle stop at some vertex, its trace would be a path with the maximum of the total preference coefficient. Now, a question we need to ask is that how one can describe the motion of the particle algebraically and how one can find a path with the maximum of the total preference coefficients once the starting vertex and the end vertex are given. To discuss these problems, we can set up an algebraic model by giving the generator set and the defining relations as follows.

Let the vertex set $V = \{v_1, v_2, \ldots, v_r\}$ be the generator set, the defining relations are given:

$$\begin{cases} v_i \cdot v_i = \sum_j w_{ij} v_j \\ v_i \cdot v_j = 0, \quad i \neq j \end{cases},$$

where preference coefficients w_{ij} and w_{ji} may be different, and $i, j = 1, 2, \ldots, r$. In this content a path with the maximum of the total preference coefficient is just a principal power of an element in the algebra; we will see this point later on.

2.2.2 Flows in a discrete space (networks)

Let us recall some basic definitions in a type of network flow theory. Let $G = (V, E)$ be a multigraph, $s, t \in V$ be two fixed vertices, and $c : \overrightarrow{E} \to N$ be a map, where N is the set of the natural numbers with zero. We call c a

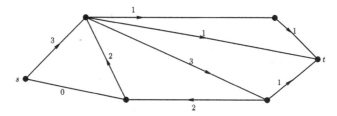

Fig. 2.2. Example of networks

capacity function on G and the tuple (G, s, t, c) a network, where $\vec{E}$ is the set of directed edges of G. Let us see an example of networks, Fig. 2.2.

Note that c is defined independently for the two directions of an edge. A function $f : \vec{E} \to R$ is a flow in the network (G, s, t, c) if it satisfies the following three conditions

(F1) $f(e, x, y) = -f(e, y, x)$, for all $(e, x, y) \in \vec{E}$ with $x \neq y$;

(F2) $f(v, V) = 0$, for all $v \in V - \{s, t\}$;

(F3) $f(\vec{e}) \leq c(\vec{e})$, for all $\vec{e} \in \vec{E}$.

Now, let us denote the capacity from vertex x to vertex y by c_{xy}, which is given by the capacity function $c(e, x, y) = c_{xy}$. We define an algebra $A(G, s, t, c)$ by generators and defining relations. The generator set is V and the defining relations are given by

$$\begin{cases} x \cdot x = \sum_y c_{xy} y \\ x \cdot y = 0, \quad x \neq y \end{cases},$$

where x and y are vertices. In the algebra $A(G, s, t, c)$, a flow is just an antisymmetric linear map. The interesting thing is that the requirement for Kirchhoff's law for a flow is automatically satisfied in the algebra.

2.2.3 Feynman graphs

Here let us recall some basic concepts in elementary particle physics. A Feynman graph [17] is a graph, each edge of which topologically represents a propagation of a free elementary particle and each vertex of which represents an interaction of elementary particles. Here, we regard a Feynman graph as an abstract object. A Feynman graph may have some extraordinary edges, called external edges, in addition to the ordinary edges, which are called internal edges. Every external edge has only one end point. A vertex is called an external vertex if at least one external edge is incident with it. Vertices other than external vertices are called internal vertices. According to the total number n of external edges, connected Feynman graphs have various names. For $n = 0$, they are called vacuum polarization graphs; $n = 1$, tadpole graphs; $n = 2$, self-energy graphs; $n = 3$, vertex graphs; $n = 4$, two-particle scattering graphs; and $n = 5$, one-particle production graphs. There are many issues in the theory of the Feynman integral that can be addressed. But here as an example to show that there exists an algebraic structure, we only mention one problem. To find some supporting properties of the Feynman integral, we need to discuss the so-called transport problem in a Feynman graph. That is, to transport given loads placed at some of vertices to the remainders as requested in such a way that when carrying a load along a edge l it does not exceed the capacity assigned to l. Similar to the previous example about the flows in a discrete space (networks), once we define an algebraic model as we did in the previous example, we will have a simple version of the original problem. So, our algebraic model can provide some insight into the theory of the Feynman integral. Below, is an example of a Feynman graph, Fig. 2.3, which yields a peculiar solution to the Landau equations and its corresponding algebra.

Denote their vertices as v_1, v_2, v_3, v_4, and two "infinite" vertices ε_1 and ε_2. The algebra corresponding to this self-energy Feynman graph is a quotient algebra whose generator set is $\{v_1, v_2, v_3, v_4, \varepsilon_1, \varepsilon_2\}$ and whose defining relations are given by

$$
\begin{aligned}
v_1^2 &= a_{12}v_2, \ v_2^2 = p\varepsilon_1, \\
v_3^2 &= a_{31}v_1 + a_{32}v_2, \\
v_4^2 &= a_{41}v_1 + a_{43}v_3 - p\varepsilon_2, \\
\varepsilon_1^2 &= \varepsilon_1, \ \varepsilon_2^2 = \varepsilon_2, \\
0 &= v_i \cdot v_j, \ i \neq j, \\
0 &= \varepsilon_1 \cdot \varepsilon_2.
\end{aligned}
$$

Here, coefficients a_{ij} and p are numbers that have physical significance.

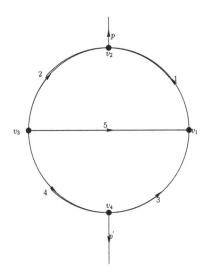

Fig. 2.3. Example of Feynman graph

2.3 Examples from Topology

2.3.1 Motions of particles in a 3-manifold

Consider a particle moving in the space (a 3-manifold M, compact or non-compact), and fix a time period t_1 to record the positions of the particle, the recorded trace of the particle is an embedded graph. There is a triangulation of the 3-manifold whose skeleton is the graph. To describe the motion, we may define

$$\begin{cases} v_i \cdot v_i = \sum_j a_{ij} v_j \\ v_i \cdot v_j = 0, \quad i \neq j, \end{cases}$$

where v_i is a vertex of the triangulation. The coefficient a_{ij} may be related to properties of the 3-manifold. For example, when the manifold carries a geometrical structure, a_{ij} may be related to the Gaussian curvature (could be negative) along the curved edge. We use these relations to define an algebra $A(M, t_1)$. This algebra will give information about the motion of the particle. When the time period of the recording is changed to t_2, we will obtain another algebra $A(M, t_2)$. Let's take an infinite sequence of time interval for recording, we will have a sequence of algebras $A(M, t_k)$. When the time interval goes to zero, we could ask what is the limit of the sequence $A(M, t_k)$. It is obvious that the sequence of these algebras reflects the properties of the manifold M. In Chapter 6, we give a different sequence of evolution algebras and an interesting conjecture related to 3-manifolds.

2.3.2 Random walks on braids with negative probabilities

In the low-dimensional topology, there is an extensive literature on the Burau representation. Jones, in his paper "Hecke algebra representation of braid groups and link polynomials" [27], offered a probabilistic interpretation of the Burau representation. We quote from this paper (with a small correction):

"For positive braids there is also a mechanical interpretation of the Burau matrix: lay the braid out flat and make it into a bowling alley with n lanes, the lanes going over each other according to the braid. If a ball travelling along a lane has probability $1 - t$ of falling off the top lane (and continuing in the lane below) at every crossing, then the (i, j) entry of the (nonreduced) Burau matrix is the probability that a ball bowled in the ith lane will end up in the jth."

Lin, Tian, and Wang, in their paper "Burau representation and random walks on string links" [28], generalized this idea to string links. Let's quote from their paper about the assignment of probability (weight) at each crossing for random walks:

(1) If we come to a positive crossing on the upper segment, the weight is $1 - t$ if we choose to jump down and t otherwise; and
(2) If we come to a negative crossing on the upper segment, the weight is $1 - \bar{t}$ if we choose to jump down and $\bar{t}$ otherwise, where $\bar{t} = t^{-1}$".

Now, we can see there are negative probabilities involved in this kind of random walks on braids. We will not go through their model here.

2.4 Examples from Probability Theory

2.4.1 Stochastic processes

Consider a stochastic process that moves through a countable set S of states. At stage n, the process decides where to go next by a random mechanism that depends only on the current state, and not on the previous history or even by the time n. These processes are called Markov chains on countable state spaces. Precisely, let X_n be a discrete-time Markov chain with state space $S = \{s_i \mid i \in \Lambda\}$, the transition probability be given by $p_{ij} = \Pr\{X_{n+1} = s_j \mid X_n = s_i\}$. Here we first consider stationary Markov chains. Then, we can reformulate such a Markov chain by an algebra. Taking the generator set as S, and the defining relations as follows

$$\begin{cases} s_i \cdot s_i = \sum_j p_{ij} s_j \\ s_i \cdot s_j = 0, \quad i \neq j \end{cases},$$

then we obtain a quotient algebra. As examples, we will study these algebras in detail in Chapter 4 of the book.

3

Evolution Algebras

As a system of abstract algebra, evolution algebras are nonassociative algebras. There is no deep structure theorem for general nonassociative algebra. However, there are deep structure theorem and classification theorem for evolution algebras because we introduce concepts of dynamical systems to evolution algebras. In this chapter, we shall introduce the foundation of the evolution algebras. Section 1 contains basic definitions and properties. Section 2 introduces evolution operators and examines related algebras, including multiplication algebras and derived Lie algebras. Section 3 introduces a norm to an evolution algebra. In Section 4, we introduce the concepts of periodicity, algebraic persistency, and algebraic transiency. In the last section, we obtain the hierarchy of an evolution algebra. For illustration, there are examples in each section.

3.1 Definitions and Basic Properties

In this section, we establish the algebraic foundation for evolution algebras. We define evolution algebras by generators and defining relations. It is notable that the generator set of an evolution algebra can serve as a basis of the algebra. We study the basic algebraic properties of evolution algebras, for example, nonassociativity, non-power-associativity, and existence of unitary elements. We also study various algebraic concepts in evolution algebras, for example, evolution subalgebras and evolution ideals. In particular, we define occurrence relations among elements of an evolution algebra and the connectedness of an evolution algebra.

3.1.1 Departure point

We define algebras in terms of generators and defining relations. The method of generators and relations is similar to the axiomatic method, where the role of axioms is played by the relations.

Let us recall the formal definition of an algebra A defined by the generators $x_1, x_2, \ldots, x_v$ and the defining relations

$$f_1 = 0, \; f_2 = 0, \; \cdots, \; f_r = 0.$$

(Both the set of generators and the set of relations, generally speaking, may be infinite. Since there is no principal difference between finite and infinite cases, we will only consider the finite cases for convenience.) We first consider a nonassociative and noncommutative free algebra $\Re$ with the set of generators $X = \{x_1, x_2, \cdots, x_v\}$ over a field K. It is necessary to point out that its elements are polynomials of noncommutative variables x_i with coefficients from K and the basis consists of bracketed words (bracketed monomials). By a bracketed word, we mean a monomial of variables $x_1, x_2, \cdots, x_v$ with brackets inserted so that the order of multiplications in the monomial is uniquely determined. In particular, all f_i are elements of this free algebra $\Re$. Then we consider the ideal I in $\Re$ generated by these elements (i.e., the smallest ideal contains these elements). The factor algebra $\Re/I$ is the algebra defined by the generators and the relations. We use notation

$$\Re/I = \langle x_1, x_2, \cdots, x_v \mid f_1, f_2, \cdots, f_r \rangle$$

for the algebra A defined by the generators $x_1, x_2, \cdots, x_v$ and the defining relations $f_1 = 0, \; f_2 = 0, \; \cdots, \; f_r = 0$.

Now let us define our evolution algebras.

Definition 1. *Let* $X = \{x_1, x_2, \cdots, x_v\}$ *be the set of generators and* $R = \{f_l = x_l^2 + \sum_{k=1}^{v} a_{lk}x_k, \; f_{ij} = x_i x_j \mid a_{lk} \in K, i \neq j, l, i, j = 1, 2, \cdots, v\}$ *be the set of defining relations, where* K *is a field, an evolution algebra is then defined by*

$$R(X) = \left\langle x_1, \cdots, x_v \mid x_l^2 + \sum_{k=1}^{v} a_{lk}x_k, \; x_i x_j, i \neq j; i, j, l \in \Lambda \right\rangle$$

where Λ *is the index set,* $\Lambda = \{1, 2, \cdots, v\}$.

Remark 1. In many practical problems, the underlying field K should be the real number field. We say an evolution algebra is real if the underlying field is the real number field R. We say an evolution algebra is nonnegative if it is real and any structural coefficient a_{jk} in defining relations is nonnegative. An evolution algebra is called Markov evolution algebra if it is nonnegative and the summation of coefficients in each defining relation is 1, $\sum_{k=1}^{v} a_{jk} = 1$, for each j. We will study Markov evolution algebras in Chapter 4.

Remark 2. There are two types of trivial evolution algebras, zero evolution algebras and nonzero trivial evolution algebras. If the defining relations are given by $x_i \cdot x_j = 0$ for all generators and any $x_i^2 = 0$, we say that the algebra generated by these generators is a zero evolution algebra. If the defining relations are given by $x_i \cdot x_j = 0$ for $i \neq j$ and $x_i \cdot x_i = k_i x_i$, where $k_i \in K$ is not a zero element, we say that the algebra generated by these generators is a nonzero trivial evolution algebra. To avoid triviality, we always assume that an evolution algebra is not a zero algebra.

To understand evolution algebras defined this way, we need to understand the properties of generators. To this end, we define a notion – the length of a bracketed word. Let $W(x_1, x_2, \cdots, x_v)$ be a bracketed word. We define the length of W, denoting it by $l(W)$, to be the sum of the number of occurrence of each generator x_i in W. Thus, for the empty word ϕ, $l(\phi) = 0$, and for any generator x_i, $l(x_i) = 1$. For example, $W = k(x_1 x_2)((x_3 x_1) x_2)$, here $l(W) = 5$, where $k \in K$. Using this notion, we can prove the following theorem.

Theorem 1. *If the set of generators X is finite, then the evolution algebra $R(X)$ is finite dimensional. Moreover, the set of generators X can serve as a basis of the algebra $R(X)$.*

Proof. We know that a general element of the evolution algebra $R(X)$ is a linear combination of reduced bracketed words. By a reduced bracketed word, we mean a bracketed word that is subject to the defining relations of $R(X)$. Therefore, if we can prove that any reduced word W can be expressed as a linear combination of generators, we can conclude that $R(X)$ has the set of generators X as its basis. Now we use induction to finish the proof.

If $l(w) = 0$, then $w = \phi$, and if $l(w) = 1$, then w must be a certain generator x_i. Furthermore, if $l(w) = 2$, w has to be x_j^2 for some generator x_j, since $x_i x_j = 0$ for two distinct generators. Since $x_j^2 + \sum_{k=1}^{v} a_{j,k} x_k = 0$, we have

$$w = x_j^2 = \sum_{k=1}^{v} -a_{j,k} x_k.$$

Now suppose that when $l(w) = n$, w can be written as a linear combination of generators. Then let us look at the case of $l(w) = n+1$. Because w here is a reduced bracketed word, the first multiplication in w must be $x_i \cdot x_i$ for a certain generator x_i; otherwise $w = \phi$. Since $x_i \cdot x_i = \sum_{k=1}^{v} -a_{i,k} x_k$, after taking the first multiplication, w will become a polynomial, each term of which has a length that is less than or equal to n. By induction, each term of the polynomial can be written as a linear combination of generators. Therefore, w can also be written as a linear combination of generators. Hence, by induction, every reduced bracketed word can be written as a linear combination of generators. Thus, the generator set X is a basis for $R(X)$.

We also need to prove that X is a linear independent set. Suppose $\sum_k a_k x_k = 0$, then multiply by x_k on both sides of the equation, we have $a_k x_k^2 = 0$. Since $x_k^2 \neq 0$, thus $a_k = 0$, for every index k (since $R(X)$ is not a zero algebra).

Actually, in the previous theorem, the restrictive condition of finiteness is not necessary, because any element of $R(X)$ is a finite linear combination of reduced bracketed words and each reduced bracketed word has a finite length whether the number of generators is finite or infinite. Therefore, we have the following two equivalent definitions for evolution algebras.

Definition 2. *Let* $S = \{x_1, x_2, \ldots, x_n, \ldots\}$ *be a countable set of letters, referred as the set of generators,* V_S *be a vector space spanned by* S *over a field* K. *We define a bilinear map* m,

$$m : \qquad V_S \times V_S \longrightarrow V_S$$

by

$$m(x_i, x_j) = 0, \quad if\ i \neq j$$
$$m(x_i, x_i) = \sum_k a_{i,k} x_k, \quad for\ any\ i$$

and bilinear extension onto $V_S \times V_S$. *Then, we call the pair* (V_S, m) *an evolution algebra.*

Or, alternatively,

Definition 3. *Let* $(A, \cdot)$ *be an algebra over a field* K. *If it admits a countable basis* $x_1, x_2, \cdots, x_n, \cdots$, *such that*

$$x_i \cdot x_j = 0, \quad if\ i \neq j$$
$$x_i \cdot x_i = \sum_k a_{i,k} x_k, \quad for\ any\ i$$

we then call this algebra an evolution algebra. We call the basis a natural basis.

Now, let us discuss several basic properties of evolution algebras. They are corollaries of the definition of an evolution algebra.

Corollary 1. *1) Evolution algebras are not associative, in general.*
2) Evolution algebras are commutative, flexible.
3) Evolution algebras are not power-associative, in general.
4) The direct sum of evolution algebras is also an evolution algebra.
5) The Kronecker product of evolution algebras is an evolution algebra.

Proof. We always work with a generator set $\{e_1, e_2, \cdots, e_n, \cdots\}$, and consider evolution algebras to be nontrivial.

1)   Generally, for some index i, $e_i \cdot e_i = \sum_j a_{ij} e_j$, there is $j \neq i$, such that $a_{ij} \neq 0$. Therefore, we have $(e_i \cdot e_i) \cdot e_j \neq 0$. But $e_i \cdot (e_i \cdot e_j) = e_i \cdot 0 = 0$. That is, $(e_i \cdot e_i) \cdot e_j \neq e_i \cdot (e_i \cdot e_j)$.

2)   For any two elements x and y in an evolution algebra, $x = \sum_i a_i e_i$ and $y = \sum_i b_i e_i$, we have

$$x \cdot y = \sum_i a_i e_i \cdot \sum_j b_j e_j = \sum_{i,j} a_i b_j e_i \cdot e_j = \sum_i a_i b_i e_i^2 = y \cdot x.$$

Therefore, any evolution algebra is commutative. Recall that an algebra is flexible if it satisfies $x(yx) = (xy)x$. It is easy to see that a commutative algebra is flexible. Therefore, any evolution algebra is flexible.

3)   Take e_i, we look at

$$(e_i \cdot e_i) \cdot (e_i \cdot e_i) = \sum_k a_{ik} e_k \cdot \sum_l a_{il} e_l = \sum_k a_{ik}^2 e_k^2$$

$$((e_i \cdot e_i) \cdot e_i) \cdot e_i = ((\sum_k a_{ik} e_k) \cdot e_i) e_i$$

$$= (a_{ii} e_i^2) \cdot e_i = (a_{ii} \sum_k a_{ik} e_k) \cdot e_i$$

$$= a_{ii}^2 e_i^2$$

generally,

$$(e_i \cdot e_i) \cdot (e_i \cdot e_i) \neq ((e_i \cdot e_i) \cdot e_i) \cdot e_i.$$

Thus, an evolution algebra is not necessarily power-associative.

4)   Consider two evolution algebras A_1, A_2 with generator sets $\{e_i \mid i \in \Lambda_1\}$ and $\{\eta_j \mid j \in \Lambda_2\}$, respectively. Then, $A_1 \oplus A_2$ has a generator set $\{e_i, \eta_j \mid i \in \Lambda_1, j \in \Lambda_2\}$, once we identify e_i with $(e_i, 0)$, η_j with $(0, \eta_j)$. Actually, this generator set is a natural basis for $A_1 \oplus A_2$. We can verify this as follows:

$$e_i \cdot e_i = \sum_k a_{ik} e_k$$
$$e_i \cdot e_j = 0, \quad \text{if } i \neq j$$
$$\eta_i \cdot \eta_i = \sum_k b_{ik} \eta_k$$
$$\eta_i \cdot \eta_j = 0, \quad \text{if } i \neq j$$
$$e_i \cdot \eta_j = (e_i, 0) \cdot (0, \eta_j) = 0.$$

Therefore $A_1 \oplus A_2$ is an evolution algebra. It is clear that the dimension of $A_1 \oplus A_2$ is the sum of the dimension of A_1 and that of A_2. The proof is similar when the number of summands of the direct sum is bigger than 2.

5)   First consider two evolution algebras A_1 and A_2 with generator sets $\{e_i \mid i \in \Lambda_1\}$ and $\{\eta_j \mid j \in \Lambda_2\}$, respectively. On the tensor product of two vector spaces A_1 and A_2, $A_1 \otimes_K A_2$, we define a multiplication in the usual way. That is, for $x_1 \otimes x_2$ and $y_1 \otimes y_2$, we define $(x_1 \otimes x_2) \cdot (y_1 \otimes y_2) = x_1 y_1 \otimes x_2 y_2$. Then, we have the Kronecker product of these two algebras. This Kronecker product is also an evolution algebra, because the generator set of

the Kronecker product is $\{e_i \otimes \eta_j \mid i \in \Lambda_1,\ j \in \Lambda_2\}$, and the defining relations are given by

$$(e_i \otimes \eta_j) \cdot (e_i \otimes \eta_j) \neq 0,$$
$$(e_i \otimes \eta_j) \cdot (e_k \otimes e_l) = 0, \text{ if } i \neq k \text{ or } j \neq l.$$

As to its dimension, we have $\dim(A_1 \otimes A_2) = \dim(A_1)\dim(A_2)$. The proof is similar when the number of factors of Kronecker product is greater than 2.

3.1.2 Existence of unity elements

For an evolution algebra A, we can use a standard construction to obtain an algebra A_1 that does contain a unity element, such that A_1 has (an isomorphic copy of) A as an ideal and A_1/A has dimension 1 over K. We take A_1 to be the set of all ordered pairs (k, x) with $k \in K$ and $x \in A$; addition and multiplication are defined by

$$(k,\ x) + (c,\ y) = (k + c,\ x + y),$$

and

$$(k,\ x) \cdot (c,\ y) = (kc,\ ky + cx + xy),$$

where $k, c \in K$, $x, y \in A$. Then A_1 is an algebra over K with unitary element $(1,\ 0)$, where 1 is the unity element of the field K and 0 is the empty element of A. The set A' of all pairs $(0,\ x)$ in A_1 with x in A is an ideal of A_1 which is isomorphic to A. For commutative Jordan algebras and alternative algebras, we know that by adjoining a unity element to them we obtain the same type of nonassociative algebras. However, in the case of evolution algebras, A_1 is no longer an evolution algebra generally. Although the subset $\{(1,0),(0,e_i) : i \in \Lambda\}$ of A_1 is a basis, and so is a generator set of algebra A_1, this subset does not satisfy the condition of generator set of an evolution algebra. The following proposition characterizes an evolution algebra with a unity element.

Proposition 1. *An evolution algebra has a unitary element if and only if it is a nonzero trivial evolution algebra.*

Proof. Let an evolution algebra A has a generator set $\{e_i \mid i \in \Lambda\}$, and $\mu = \sum_i a_i e_i$ be a unity element. We then have $\mu e_i = e_i$ for each $i \in \Lambda$. That is,

$$e_i = \left(\sum_j a_j e_j\right) e_i = a_i e_i^2 = a_i \sum_j a_{ij} e_j.$$

We have to have $a_i a_{ii} = 1$ and $a_{ij} = 0$ if $i \neq j$. That means A must be a nonzero trivial evolution algebra, and the unity element is given by $\mu = \sum_i \frac{1}{a_{ii}} e_i$. On the other hand, if A is a nonzero trivial evolution algebra, it is easy to check that there is a unity element, which is given by μ.

3.1.3 Basic definitions

We need some more basic definitions: evolution subalgebras, evolution ideals, principal powers, plenary powers, and simple evolution algebras. Now, let's define them.

Definition 4. *1) Let A be an evolution algebra, and A_1 be a subspace of A. If A_1 has a natural basis $\{e_i \mid i \in \Lambda_1\}$, which can be extended to a natural basis $\{e_j \mid j \in \Lambda\}$ of A, we call A_1 an evolution subalgebra, where Λ_1 and Λ are index sets and Λ_1 is a subset of Λ.*

2) Let A be an evolution algebra, and I be an evolution subalgebra of A. If $AI \subseteq I$, we call I an evolution ideal.

3) Let A and B be evolution algebras, we say a linear homomorphism f from A to B is an evolution homomorphism, if f is an algebraic map and for a natural basis $\{e_i \mid i \in \Lambda\}$ of A, $\{f(e_i) \mid i \in \Lambda\}$ spans an evolution subalgebra of B. Furthermore, if an evolution homomorphism is one to one and onto, it is an evolution isomorphism.

4) Let A be a commutative algebra, we define principal powers of $a \in A$ as follows:

$$a^2 = a \cdot a$$
$$a^3 = a^2 \cdot a$$
$$\cdots\cdots$$
$$a^n = a^{n-1} \cdot a;$$

and plenary powers of $a \in A$ as follows:

$$a^{[1]} = a^{(2)} = a \cdot a$$
$$a^{[2]} = a^{(2^2)} = a^{(2)} \cdot a^{(2)}$$
$$a^{[3]} = a^{(2^3)} = a^{(4)} \cdot a^{(4)}$$
$$\cdots\cdots\cdots$$
$$a^{[n]} = a^{(2^n)} = a^{(2^{n-1})} \cdot a^{(2^{n-1})}.$$

For convenience, we denote $a^{[0]} = a$.

Then, we have a property

$$\left(a^{[n]}\right)^{[m]} = a^{[n+m]},$$

where n and m are positive integers. The proof of this property can be obtained by counting the number of a that contains in the mth plenary power of $a^{[n]}$, therefore

$$\left(a^{[n]}\right)^{[m]} = \left(a^{(2^n)}\right)^{(2^m)} = a^{(2^n 2^m)} = a^{(2^{n+m})} = a^{[n+m]}.$$

5) We say an evolution algebra E is connected if E can not be decomposed into a direct sum of two proper evolution subalgebras.

6) An evolution algebra E is simple if it has no proper evolution ideal.

7) An evolution algebra E is irreducible if it has no proper subalgebra.

Natural bases of evolution algebras play a privileged role among all other bases, since the generators represent alleles in genetics and states generally in other problems. Importantly, natural bases are privileged for mathematical reasons, too. The following example illustrates this point.

Example 1. Let E be an evolution algebra with basis e_1, e_2, e_3 and multiplication defined by $e_1e_1 = e_1 + e_2$, $e_2e_2 = -e_1 - e_2$, $e_3e_3 = -e_2 + e_3$. Let $u_1 = e_1 + e_2$, $u_2 = e_1 + e_3$. Then $(\alpha u_1 + \beta u_2)(\gamma u_1 + \delta u_2) = \alpha\gamma u_1^2 + (\alpha\delta + \beta\gamma)u_1u_2 + \beta\delta u_2^2 = (\alpha\delta + \beta\gamma)u_1 + \beta\delta u_2$. Hence, $F = Ku_1 + Ku_2$ is a subalgebra of E. However, F is not an evolution subalgebra.

Let v_1, v_2 be a basis of F. Then $v_1 = \alpha u_1 + \beta u_2$, $v_2 = \gamma u_1 + \delta u_2$ for some $\alpha, \beta, \gamma, \delta \in K$ such that $D = \alpha\delta - \beta\gamma \neq 0$. By the above calculation, $v_1v_2 = (\alpha\delta + \beta\gamma)u_1 + \beta\delta u_2$. Assume that $v_1v_2 = 0$. Then $\beta\delta = 0$ and $\alpha\delta + \beta\gamma = 0$. If $\beta = 0$, we have $\alpha\delta = 0$. Then, $D = 0$, a contradiction. If $\delta = 0$, we reach the same contradiction. Hence $v_1v_2 \neq 0$, and F is not an evolution subalgebra.

We have just seen that evolution algebras are not closed under subalgebras. This is one reason we define these new notions, such as evolution subalgebras. We shall see the relations between these concepts in next subsection.

3.1.4 Ideals of an evolution algebra

Classically, an ideal I in an algebra A is first a subalgebra, and then it satisfies $AI \subseteq I$ and $IA \subseteq I$. In the setting of evolution algebras, an evolution ideal is first an evolution subalgebra. However, the conditions for evolution subalgebras seem enough for evolution ideals. We have the following property.

Proposition 2. *Any evolution subalgebra is an evolution ideal.*

Proof. Let E_1 be an evolution subalgebra of E, then E_1 has a generator set $\{e_i \mid i \in \Lambda_1\}$ that can be extended to a generator set of E, $\{e_i \mid i \in \Lambda\}$, where Λ_1 is a subset of Λ. For $x \in E_1$, and $y \in E$, we write $x = \sum_{i \in \Lambda_1} x_i e_i$ and $y = \sum_{i \in \Lambda} y_i e_i$, where $x_i, y_i \in K$, we then have the product $xy = \sum_{i \in \Lambda_1} x_i y_i e_i^2 \in E_1$. Therefore, $E_1 E \subseteq E_1$. Since E is a commutative algebra, E_1 is a two-sided ideal.

This property makes the concept of evolution ideals superfluous. We will use the notion, evolution ideals, as an equivalent concept of evolution subalgebras. As we know, a simple algebra does not have a proper ideal. And an evolution algebra is irreducible if it does not have a proper subalgebra. So, from the above proposition, an irreducible evolution algebra is a simple evolution algebra, and a simple evolution algebra is an irreducible evolution algebra. They are, actually, the same concepts in evolution algebras. As in general algebra theory, if an evolution algebra can be written as a direct sum of evolution subalgebras, we call it a semisimple evolution algebra. Then we have the following corollary.

Corollary 2. *1) A semisimple evolution algebra is not connected.*
2) A simple evolution algebra is connected.

3.1.5 Quotients of an evolution algebra

To study structures of evolution algebras, particularly, hierarchies of evolution algebras, quotients of evolution algebras should be studied. Let E_1 be an evolution ideal of an evolution algebra E, then the quotient algebra $\overline{E} = E/E_1$ consists of all cosets $\overline{x} = x + E_1$ with the induced operations $k\overline{x} = \overline{kx}$, $\overline{x} + \overline{y} = \overline{x + y}$, $\overline{x} \cdot \overline{y} = \overline{xy}$. We can easily verify that $\overline{E}$ is an evolution algebra. The canonical map $\pi : x \mapsto \overline{x}$ of E onto $\overline{E}$ is an evolution homomorphism with the kernel E_1.

Lemma 1. *Let η_1, η_2, $\cdots$, η_m be elements of an evolution algebra E with dimension n, and satisfies $\eta_i \eta_j = 0$ when $i \neq j$. If some of these elements form a basis of E, then there are $(m - n)$ zeroes in this sequence.*

Proof. Suppose η_1, η_2, $\cdots$, η_n form a natural basis of E. Then, η_{n+k}, $1 \leq k \leq (m - n)$, can be expressed as a linear combination of η_i, $1 \leq i \leq n$. That is, $\eta_{n+k} = \sum_{i=1}^{n} a_i \eta_i$. Multiplying by η_i on both sides of this equation, we have $\eta_{n+k}\eta_i = a_i \eta_i^2 = 0$; then, $a_i = 0$, for each i, $1 \leq i \leq n$. Therefore, $\eta_{n+k} = 0$, where $1 \leq k \leq m - n$.

Theorem 2. *Let E_1 and E_2 be evolution algebras, and $f : E_1 \longrightarrow E_2$ be an evolution algebraic homomorphism. Then, $K = kernel(f)$ is an evolution subalgebra of E_1, and E_1/K is isomorphic to E_2 if f is surjective. Or, E_1/K is isomorphic to $f(E_1)$.*

Proof. Let e_1, e_2, $\cdots$, e_m be a natural basis of E_1, by the definition of evolution algebra homomorphism, $f(e_1)$, $f(e_2)$, $\cdots$, $f(e_m)$ span an evolution subalgebra of E_2; denote this subalgebra by B. When $dim(B) = m$, it is easy to see that $K = kernel(f) = 0$. K is the zero subalgebra. When $dim(B) = n < m$, we will prove $dim(K) = m - n$. For $i \neq j$, $f(e_i)f(e_j) = f(e_ie_j) = 0$, and some of $f(e_i)$s form a natural basis of the image of E_1, which is an evolution subalgebra of E_2. By the Lemma 1, there are $m - n$ zeroes; let's say $f(e_{n+1}) = 0$, $\cdots$, $f(e_m) = 0$. That means, $e_{n+1}, \cdots, e_m \in K$. Actually, they span an evolution subalgebra, which is the kernel K of f with dimension $m - n$.

Set a map

$$\overline{f} : E_1/K \longrightarrow f(E_1)$$

by

$$x + K \longmapsto f(x).$$

It is not hard to see that $\overline{f}$ is an isomorphic.

We may conclude that an evolution algebra can be homomorphic and can only be homomorphic to its quotients. We will study the automorphism group of an evolution algebra in the next section.

3.1.6 Occurrence relations

When an element in a basis is viewed as an allele in genetics, or a state in stochastic processes, we are most interested in the following questions: when does the allele e_i give rise to the allele e_j? when does a state appear in the next step of the process? To address this question, we introduce a notion, occurrence relations.

Let E be an evolution algebra with the generator set $\{e_1, e_2, \cdots, e_v\}$. We say e_i occurs in $x \in E$, if the coefficient $\alpha_i \in K$ is nonzero in $x = \sum_{j=1}^{v} \alpha_j e_j$. When e_i occurs in x, we write $e_i \prec x$.

It is not hard to see that if $e_i \prec e_i^{[n]}$, then $\langle e_i \rangle \subseteq \langle e_i \rangle$, where $\langle x \rangle$ means the evolution subalgebra generated by x.

When we work on nonnegative evolution algebras, we can obtain a type of partial order among elements.

Lemma 2. *Let E be a nonnegative evolution algebra. Then for every $x, y \in E^+$, and $n \geq 0$, there is $z \in E^+$, such that $(x + y)^{[n]} = x^{[n]} + z$, where $E^+ = \sum \alpha_i e_i; \alpha_i \geq 0$.*

Proof. We prove the lemma by induction on n. We have $(x + y)^{[0]} = x^{[0]} + y$, and it suffices to set $z = y$. Also, $(x + y)^{[1]} = x^{[1]} + 2xy + y^2$. Since E^+ is closed under addition, multiplication, and multiplication by positive scalars, $z = 2xy + y^2$ belongs to E^+.

Assume the claim is true for $n > 1$. In particular, give $x, y \in E^+$, let $w \in E^+$ such that $(x + y)^{[n]} = x^{[n]} + w$. Then $(x + y)^{[n+1]} = (x^{[n]} + w)^{[1]} = (x^{[n]})^{[1]} + z = x^{[n+1]} + z$ for some $z \in E^+$.

Proposition 3. *Let E be a nonnegative evolution algebra. When $e_i \prec e_j^{[n]}$ and $e_j \prec e_k^{[m]}$, then $e_i \prec e_k^{[n+m]}$*

Proof. We have $e_k^{[m]} = \alpha_j e_j + y$ for some $\alpha_j \neq 0$ and $y \in E$, such that e_j does not occur in y. We also have $\alpha_j > 0$ and $y \in E^+$. By Lemma 2, $e_k^{[n+m]} = (e_k^{[m]})^{[n]} = (\alpha_j e_j + y)^{[n]} = (\alpha_j e_j)^{[n]} + z = \alpha_j^{(2^n)} e_j^{[n]} + z$ for some $z \in E^+$. Now, $e_j^{[n]} = \beta_i e_i + v$ for some $\beta_i > 0$ and $v \in E$ that e_i does not occur in v. We therefore conclude that $e_i \prec e_k^{[n+m]}$.

We can have a type of partial order relation among the generators of an evolution algebra E. Let e_i and e_j be any two generators of E, if e_i occurs in a plenary power of e_j, for example, e_i occurs in $e_j^{[n]}$, we then set $e_i < e_j$, or just $e_i \prec e_j^{[n]}$. This relation is a partial order in the following sense.

(1) $e_i \prec e_i^{[0]}$, for any generator of E.
(2) If $e_i \prec e_j^{[n]}$ and $e_j \prec e_i^{[m]}$, then we say that e_i and e_j intercommunicate. Generally, e_i and e_j are not necessarily the same, but the evolution subalgebra generated by e_i and the one by e_j are the same.
(3) If $e_i \prec e_j^{[n]}$ and $e_j \prec e_k^{[m]}$, then $e_i \prec e_k^{[n+m]}$. This is Proposition 3.

3.1.7 Several interesting identities

At the end of this section, let us give several interesting formulae, they are identities.

Proposition 4. *1) Let $\{e_i \mid i \in \Lambda\}$ be a natural basis of an evolution algebra A, then $\{e_i^2 \mid i \in \Lambda\}$ generates a subalgebra A.*
2) Let $\{e_i \mid i \in \Lambda\}$ be a natural basis of an evolution algebra A, then we have the following identities:

$$e_i^m = a_{ii}^{m-2} e_i^2, \qquad \forall\, i \in \Lambda, \qquad \forall\, m \geq 2$$
$$e_i^2 \cdot e_j = a_{ij} e_j^2, \qquad \forall\, i,\, j \in \Lambda,$$
$$(e_i^m)^2 = a_{ii}^{2m-4} e_i^{(4)}, \quad \forall\, i \in \Lambda, \qquad \forall\, m \geq 2$$
$$e_i^4 \cdot e_i^4 = a_{ii}^4 e_i^{(2)}, \qquad \forall\, i,\, j \in \Lambda,$$

where a_{ij}'s are structural constants of A.
3) Let $\{e_i \mid i \in \Lambda\}$ be a natural basis of an evolution algebra, then, for any finite subset Λ_0 of the index set Λ, we have

$$\left(\sum_{j \in \Lambda_0} e_j \right)^2 = \sum_{j \in \Lambda_0} e_j^2.$$

Proof. 1) Since $\{e_i \mid i \in \Lambda\}$ be a generator set, so

$$e_i^2 = \sum_k a_{ik} e_k,$$
$$e_i^2 \cdot e_i^2 = \sum_k a_{ik} e_k \cdot \sum_l a_{il} e_l = \sum_{l,\,k} a_{ik} a_{il} e_k \cdot e_l = \sum_k a_{ik}^2 e_k^2,$$
$$e_i^2 \cdot e_j^2 = \sum_k a_{ik} e_k \cdot \sum_l a_{jl} e_l = \sum_{l,\,k} a_{ik} a_{jk} e_k^2.$$

Thus, any product of linear combinations of e_i^2 can still be written as a linear combination of e_i^2. This means that $\{e_i^2 \mid i \in \Lambda\}$ generates a subalgebra of A.

2) Since

$$e_i^2 = \sum_k a_{ik} e_k,$$

$$e_i^3 = e_i^2 \cdot e_i = \left(\sum_k a_{ik} e_k\right) \cdot e_i = a_{ii} e_i^2.$$

If $e_i^{m-1} = a_{ii}^{m-3} e_i^2$, for any integer $m > 2$, then

$$e_i^m = e_i^{m-1} \cdot e_i = a_{ii}^{m-3} e_i^2 \cdot e_i = a_{ii}^{m-3}\left(\sum_k a_{ik} e_k\right) \cdot e_i = a_{ii}^{m-2} e_i^2.$$

By induction, we got the first formula.

As to the second formula, we have

$$e_i^2 \cdot e_j = \left(\sum_k a_{ik} e_k\right) \cdot e_j = a_{ij} e_j^2.$$

As to the third formula, we see

$$(e_i^m)^2 = e_i^m \cdot e_i^m = a_{ii}^{2m-4} e_i^2 \cdot e_i^2 = a_{ii}^{2m-4} e_i^{(4)}.$$

Taking $m = 4$, we have

$$e_i^4 \cdot e_i^4 = a_{ii}^4 e_i^2 \cdot e_i^2 = a_{ii}^4 e_i^{(4)}.$$

3) By directly computing, we have

$$\left(\sum_{j \in \Lambda_0} e_j\right)^2 = \sum_{j \in \Lambda_0} e_j \cdot \sum_{i \in \Lambda_0} e_i = \sum_{i,\, j \in \Lambda_0} e_i \cdot e_j = \sum_{j \in \Lambda_0} e_j^2.$$

3.2 Evolution Operators and Multiplication Algebras

Traditionally, in the study of nonassociative algebras, one usually studies the associative multiplication algebra of a nonassociative algebra and its derived Lie algebra to try to understand the nonassociative algebra. In this section, we also study the multiplication algebra of an evolution algebra and conclude that any evolution algebra is centroidal. We characterize the automorphism group of an evolution algebra and its derived Lie algebra. Moreover, from the viewpoint of dynamics, we introduce the evolution operator for an evolution algebra. This evolution operator will reveal the dynamic information of an evolution algebra. Because we work with a generator set of an evolution algebra, it is also necessary for us to study the change of generator set, or transformations of natural bases.

3.2.1 Evolution operators

Definition 5. *Let E be an evolution algebra with a generator set $\{e_i \mid i \in \Lambda\}$. We define a K-linear map L to be*

$$L : E \longrightarrow E$$
$$e_i \mapsto e_i^2 \ \forall \ i \in \Lambda$$

then linear extension onto E.

Consider L as a linear transformation, ignoring the algebraic structure of E, then under a natural basis (the generator set), we can have the matrix representation of the evolution operator L. Since

$$L(e_i) = e_i^2 = \sum_k a_{ki} e_k \qquad \forall i \in \Lambda,$$

then we have

$$\begin{pmatrix} a_{11} & a_{12} & \cdots & a_{1n} & \cdots \\ a_{21} & a_{22} & \cdots & a_{2n} & \cdots \\ \vdots & \vdots & \vdots & \vdots & \vdots \\ a_{n1} & a_{n2} & \cdots & a_{nn} & \cdots \\ \vdots & \vdots & \vdots & \vdots & \vdots \end{pmatrix}.$$

If E is a finite dimensional algebra, this matrix will be of finite size. An evolution operator, not being an algebraic map though, can reveal dynamical properties of the evolution algebra, as we will see later on.

Alternatively, by using a formal notation $\theta = \sum_{i \in \Lambda} e_i$, no matter whether Λ is finite or infinite, we can define L as follows:

$$L(x) = \theta \cdot x = \left(\sum_{i \in \Lambda} e_i\right) \cdot x,$$

for any $x \in E$. According to the distributive law of product to addition in algebra E, L is a linear map. Because

$$L(e_i) = \left(\sum_{i \in \Lambda} e_i\right) \cdot e_i = e_i^2, \qquad \forall i \in \Lambda,$$

this definition for an evolution operator is the same as the previous one. We do not feel uncomfortable about the notation $\theta = \sum_{i \in \Lambda} e_i$, when Λ is infinite, since the product $\left(\sum_{i \in \Lambda} e_i\right) \cdot x$ is always finite. We may call this θ a universal element.

Now, we state a theorem that will be used to get the equilibrium state or a fixed point of the evolution of an evolution algebra.

Theorem 3. *If E_0 is an evolution subalgebra of an evolution algebra E, then the evolution operator L of E leaves E_0 invariant.*

Proof. Let $\{e_i \mid i \in \Lambda_0\}$ be a natural basis of E_0, and $\{e_i \mid i \in \Lambda\}$ be its extension to a natural basis of E, where $\Lambda_0 \subset \Lambda$. Given $x \in E_0$, then $x = \sum_{i \in \Lambda_0} c_i e_i$, and the action of the evolution operator is

$$L(x) = \sum_{i \in \Lambda_0} c_i e_i^2 = \sum_{i \in \Lambda_0, \ k \in \Lambda_0} c_i a_{ki} e_k,$$

since E_0 is a subalgebra. Therefore, $L(x) \in E_0$, then $L(E_0) \subset E_0$. Furthermore, $L^n(E_0) \subset E_0$, for any positive integer n.

3.2.2 Changes of generator sets (Transformations of natural bases)

Let $\{e_i \mid i \in \Lambda\}$ and $\{\eta_j \mid j \in \Lambda\}$ be two generator sets (natural bases) for an evolution algebra E. Suppose the transformation between them is given by $e_i = \sum_k a_{ki} \eta_k$ or $\eta_i = \sum_k b_{ki} e_k$. And suppose the defining relations are $e_i \cdot e_j = 0$ if $i \neq j$, $e_i^2 = \sum_k p_{ki} e_k$, and $\eta_i \cdot \eta_j = 0$ if $i \neq j$, $\eta_i^2 = \sum_k q_{ki} \eta_k$, $i, j \in \Lambda$, respectively. Then, we have

$$e_i \cdot e_j = \left(\sum_k a_{ki} \eta_k \right) \cdot \left(\sum_k a_{kj} \eta_k \right)$$
$$= \sum_k a_{ki} a_{kj} \eta_k^2 = \sum_{v,k} a_{ki} a_{kj} q_{vk} \eta_v$$
$$= \sum_v \sum_k q_{vk} a_{ki} a_{kj} \eta_v = 0.$$

Since each component coefficient of zero vector must be 0, we get $\sum_k q_{vk} a_{ki} a_{kj} = 0$ for $v \in \Lambda$ and $i \neq j$. Similarly, from

$$e_i \cdot e_i = \left(\sum_k a_{ki} \eta_k \right)^2 = \sum_k a_{ki}^2 \eta_k^2$$
$$= \sum_{v,k} a_{ki}^2 q_{vk} \eta_v = \sum_{v,k,u} a_{ki}^2 q_{vk} b_{uv} e_u$$
$$= \sum_u p_{ui} e_u,$$

we get $p_{ui} = \sum_{v,k} b_{uv} q_{vk} a_{ki}^2$. Thus, summarizing all these information together, we have

$$A^{-1} Q A^{(2)} = P,$$
$$Q(A * A) = 0,$$

where $A = (a_{ij})$, $Q = (q_{ij})$, $P = (p_{ij})$, $A^{(2)} = (a_{ij}^2)$ and "$*$" of two matrices is defined as follows.

Let $A = (a_{ij})$ and $B = (b_{ij})$ be two $n \times n$ matrices, then $A * B = (c_{ij}^k)$ is a matrix with size $n \times \frac{n(n-1)}{2}$, where $c_{ij}^k = a_{ki} \cdot b_{kj}$ for pairs (i,j) with $i < j$, the rows are indexed by k and the columns indexed by pairs (i,j) with the lexicographical order.

We can also use B to describe the above condition

$$B^{-1}PB^{(2)} = Q,$$
$$P(B * B) = 0,$$

where $BA = AB = I$.

3.2.3 "Rigidness" of generator sets of an evolution algebra

By "rigidness," we mean that an evolution operator is specified by a generator set. Let us illustrate this point in the following way. Given a generator set $\{e_i \mid i \in \Lambda\}$, we have an evolution operator, denoted by L_e. When the generator set is changed to $\{\eta_j \mid j \in \Lambda\}$, we also have an evolution operator, denoted by L_η. Since a generator set is also a natural basis in evolution algebras, it might be expected that L_e and L_η, as linear maps, should be the same. However, they are different, unless additional conditions are imposed. Therefore, an evolution operator is not just a linear map. It is a map related to a specific generator set. This property is very useful to study the dynamic behavior of an algebra, because a multiplication in an algebra is viewed as a dynamical step. In the following lemma, we describe an additional condition about transformations of natural bases that guarantee L_e and L_η will be the same linear map.

Lemma 3. *L_e and L_η are the same invertible linear map if and only if the generator sets $\{e_i \mid i \in \Lambda\}$ and $\{\eta_j \mid j \in \Lambda\}$ are the same, or if one can be obtained from the other by a permutation.*

Proof. Here we use the same notations as those used in the previous subsection. The matrix representation of L_η is Q under the generator set $\{\eta_j \mid j \in \Lambda\}$, and

$$L_\eta(e_1, e_2, \cdots, e_n) = L_\eta(\eta_1, \eta_2, \cdots, \eta_n) A$$
$$= (\eta_1, \eta_2, \cdots, \eta_n) QA$$
$$= (e_1, e_2, \cdots, e_n) A^{-1}QA.$$

Thus, the matrix representation of L_η is $A^{-1}QA$ under the generator set $\{e_i \mid i \in \Lambda\}$. But as we know, the matrix representation of L_e is P under the natural basis $\{e_i \mid i \in \Lambda\}$. Therefore, $P = A^{-1}QA$, if L_η and L_e can be taken as the same linear maps. From the previous subsection, we know $A^{-1}QA^{(2)} = P$, so we have $A^{-1}QA = A^{-1}QA^{(2)}$. Since L_η is invertible, we then have $A = A^{(2)}$. Similarly, we have $B = B^{(2)}$. Since $a_{ij} = a_{ij}^2$, a_{ij} must be 1 or 0 and b_{ij} must also be 1 or 0, then we can prove A can only be a permutation matrix as follows:

$$\begin{pmatrix} a_{11} & a_{12} & \cdots & a_{1n} \\ a_{21} & a_{22} & \cdots & a_{2n} \\ \cdots & \cdots & \cdots & \cdots \\ a_{n1} & a_{n2} & \cdots & a_{nn} \end{pmatrix} \begin{pmatrix} b_{11} & b_{12} & \cdots & b_{1n} \\ b_{21} & b_{22} & \cdots & b_{2n} \\ \cdots & \cdots & \cdots & \cdots \\ b_{n1} & b_{n2} & \cdots & b_{nn} \end{pmatrix} = \begin{pmatrix} 1 & \cdots & 0 \\ \cdots & \ddots & \cdots \\ 0 & \cdots & 1 \end{pmatrix}$$

Without loss of generality, suppose that $a_{11} \neq 0$, $a_{12} \neq 0$, and $a_{1k} = 0$ for $k \geq 3$. Then we have $a_{11}b_{11} + a_{12}b_{21} = 1$. Thus, we have either $b_{11} \neq 0$ or $b_{21} \neq 0$. But only one of these two entries can be nonzero, otherwise $a_{11}b_{11} + a_{12}b_{21} = 2$. Now, suppose $b_{21} \neq 0$, and $b_{11} = 0$, then $a_{11}b_{12} + a_{12}b_{22} = 0$, then we must have $b_{12} = 0$; and by $a_{11}b_{13} + a_{12}b_{23} = 0$, we have $b_{13} = 0$; inductively, $b_{1j} = 0$, $j = 2, 3, \cdots$. This means $b_{11} = b_{12} = \cdots = b_{1n} = 0$. This contradicts the nonsingularity of B. If we suppose $b_{11} \neq 0$, and $b_{21} = 0$, similarly we get $b_{21} = b_{22} = \cdots = b_{2n} = \cdots = 0$. That is a contradiction. Therefore, every row of A can only have one entry that is not zero. Similarly, we can prove that every column of A can only have one entry that is nonzero. Therefore, A is a permutation matrix.

3.2.4 The automorphism group of an evolution algebra

Given an evolution algebra E, it is important to know how many generator sets E can have. To study this problem, we need to study the automorphism group of an evolution algebra.

Proposition 5. *Let g be an automorphism of an evolution algebra E with a generator set $\{e_i \mid i \in \Lambda\}$, then $G^{-1}PG^{(2)} = P$ and $P(G * G) = 0$, where G and P are the matrix representations of g and L respectively.*

Proof. Write $g(e_i) = \sum_k g_{ki}e_k$ and $G = (g_{ij})$. For $i \neq j$, we have

$$g(e_i \cdot e_j) = 0$$
$$= g(e_i)g(e_j)$$
$$= \sum_k g_{ki}e_k \cdot \sum_k g_{kj}e_k$$
$$= \sum_k g_{ki}g_{kj}e_k^2$$
$$= \sum_{k,v} p_{vk}g_{ki}g_{kj}e_v.$$

So we have $\sum_k p_{vk}g_{ki}g_{kj} = 0$, for each v. That is $P(G * G) = 0$. For $i = i$,

$$g\left(e_i \cdot e_i\right) = g\left(e_i\right) g\left(e_i\right)$$

$$= \sum_k g_{ki}^2 e_k^2$$

$$= \sum_{k,j} g_{ki}^2 p_{jk} e_j$$

$$= \sum_{k,j} p_{ki} g_{jk} e_j.$$

Thus, we have $\sum_k p_{jk} g_{ki}^2 = \sum_k g_{jk} p_{ki}$. That is $PG^{(2)} = GP$, thus $G^{-1} PG^{(2)} = P$.

Therefore, we can characterize the automorphism group of E as

$$Auto\left(E\right) = \left\{ G \mid G^{-1} PG^{(2)} = P, \text{ and } P\left(G * G\right) = 0 \right\}.$$

We can use the automorphism group to give a description of the collection of all generator sets. We write it as a corollary.

Corollary 3. Let $B = e_i : i \in \Lambda$ be a generator set of an evolution algebra E. Then the family $g(B) : g \in Auto(E)$ is the collection of all different generator sets of E.

3.2.5 The multiplication algebra of an evolution algebra

Let E be an algebra, denote L_a and R_a as the operators of the left and right multiplication by the element a respectively:

$$L_a : x \mapsto a \cdot x$$
$$R_a : x \mapsto x \cdot a.$$

The subalgebra of the full matrix algebra $Hom\left(E, E\right)$ of the endomorphisms of the linear space E, generated by all the operators L_a, $a \in E$, is called the operator algebra of left multiplication of the algebra E, denoted by $L(E)$. The operator algebra of right multiplication $R(E)$ of the algebra E is defined analogously. The subalgebra of $Hom\left(E, E\right)$ generated by all the operators L_a, R_a, $a \in E$ is called the multiplication algebra of the algebra E, denoted by $M(E)$, which is actually the enveloping algebra of all operators L_a, R_a, $a \in E$.

Corollary 4. If E is an evolution algebra, $L(E) = R(E) = M(E)$ is an associative algebra with a unit.

Proof. Since E is commutative, it is obvious.

Corollary 5. If E is an evolution algebra with a natural basis $\{e_i \mid i \in \Lambda\}$, then $\{L_i \mid i \in \Lambda\}$ spans a linear space, denoted by $span(L, E)$, which is the set of all the operators of left (right) multiplication, where $L_i = L_{e_i}$. The vector space $span(L, E)$ and E have the same dimension. Generally, we also have $\dim(E) < \dim(L(E))$ if $\dim E^2 \neq 1$.

Proof. For any operator of left multiplication L_x, we can write $x = \sum_i a_i e_i$ uniquely, then by the linearity of multiplication in E, $L_x = \sum_i a_i L_i$. If

$$L_x = L_y,$$

for $y = \sum_i b_i e_i$, then

$$L_x(e_k) = L_y(e_k), \text{ and, } \left(\sum_i a_i e_i\right) \cdot e_k = \left(\sum_i b_i e_i\right) \cdot e_k.$$

Thus,

$$a_k e_k^2 = b_k e_k^2,$$
$$(a_k - b_k) e_k^2 = 0,$$
$$(a_k - b_k) \sum_i p_{ki} e_i = 0.$$

Since E is a nontrivial algebra, there is j, $p_{kj} \neq 0$, and $(a_k - b_k) p_{kj} e_j = 0$, thus $a_k - b_k = 0$ for each k. Therefore $x = y$. This means that $x \mapsto L_x$ is an injection. So the linear space that is spanned by all operators of left multiplication can be spanned by the set $\{L_i | i \in \Lambda\}$. Moreover the set $\{L_i | i \in \Lambda\}$ is a basis for $span(L, E)$. However, since the algebra E is not associative, $x \mapsto L_x$ is not an algebraic map from E to $L(E)$. Generally, $\{L_i | i \in \Lambda\}$ is not a basis for $L(E)$. Since $\dim E^2 > 1$, there are different generators e_i and e_j whose square vectors e_i^2 and e_j^2 are not parallel to each other. For the sake of simplicity, we denote them as e_1 and e_2. We claim that $L_2 \circ L_1$ can not be represented by a linear combination of L_i, $i \in \Lambda$. Suppose $L_2 \circ L_1 = \sum_i a_i L_i$, then

$$L_2 \circ L_1(e_k) = \left(\sum_i a_i L_i\right)(e_k);$$

$$k \neq 1, \quad 0 = a_k e_k^2, \quad a_k = 0;$$
$$k = 1, \quad L_2(e_1^2) = a_1 e_1^2, \quad p_{12} e_2^2 = a_1 e_1^2;$$

so

$$p_{12} p_{2k} = a_1 p_{1k}, \ \forall k.$$

If a_1 was not zero, $p_{1k} = \frac{p_{12}}{a_1} p_{2k}, \forall k$, but it is not possible since e_1^2 and e_2^2 are not parallel. Therefore, $L(E)$ can not be spanned by $\{L_i \mid i \in \Lambda\}$.

3.2.6 The derived Lie algebra of an evolution algebra

As for any algebra, the subspace $Der(E)$ of derivations of an evolution E is a Lie algebra. Here, let us characterize an element that belongs to the $Der(E)$. Let $\{e_i \mid i \in \Lambda\}$ be a generator set of E, $D \in Der(E)$, and suppose

$D(e_i) = \sum_k d_{ki} e_k$ for $i \in \Lambda$. By the definition of derivation $D(xy) = D(x)y + xD(y)$, we have

$$D(e_i e_j) = D(e_i)e_j + e_i D(e_j)$$

$$= \left(\sum_k d_{ki} e_k \right) e_j + e_i \left(\sum_k d_{kj} e_k \right)$$

$$= d_{ji} e_j^2 + d_{ij} e_i^2$$

$$= d_{ji} \sum_k p_{kj} e_k + d_{ij} \sum_k p_{ki} e_k$$

$$= \sum_k (d_{ji} p_{kj} + d_{ij} p_{ki}) e_k$$

$$= 0,$$

so, for $i \neq j$, $p_{kj} d_{ji} + p_{ki} d_{ij} = 0$, $i \in \Lambda$. We also have

$$D(e_i^2) = D\left(\sum_k p_{ki} e_k \right)$$

$$= \sum_k p_{ki} D(e_k)$$

$$= \sum_{j,k} p_{ki} d_{jk} e_j$$

$$= 2 \sum_j d_{ii} p_{ji} e_j,$$

so, we get for any $i, j \in \Lambda$, $2p_{ji} d_{ii} = \sum_k p_{ki} d_{jk}$. Therefore, we have

$$Der(E) = \left\{ D \in End(E) \mid p_{kj} d_{ji} + p_{ki} d_{ij} = 0, \text{ for } i \neq j; 2p_{ji} d_{ii} \right.$$

$$\left. = \sum_k p_{ki} d_{jk} \right\}.$$

3.2.7 The centroid of an evolution algebra

We recall that the centroid $\Gamma(E)$ of an algebra E is the set of all linear transformations $T \in Hom(E, E)$ that commute with all left and right multiplication operators

$$TL_x = L_x T, \quad TR_y = R_y T, \quad \text{for all } x, y \in E.$$

Or, the centroid centralizes the multiplication algebra $M(E)$. That is

$$\Gamma(E) = Cent_{Hom(E,E)}(M(E)).$$

Theorem 4. *Any evolution algebra is centroidal.*

Proof. Let T be an element of the centroid $\Gamma(E)$. Suppose $T(e_i) = \sum_k t_{ki} e_k$, for $i \neq j$, we have

$$
\begin{aligned}
TL_{e_j}(e_i) = T(e_j e_i) &= 0 \\
&= L_{e_j} T(e_i) \\
&= e_j \left(\sum_k t_{ki} e_k \right) \\
&= t_{ji} e_j^2 = t_{ji} \sum_k p_{kj} e_k,
\end{aligned}
$$

thus, $t_{ij} = 0$. Then, look at

$$
\begin{aligned}
TL_{e_i}(e_i) = T(e_i^2) &= T\left(\sum_k p_{ki} e_k \right) \\
&= \sum_k p_{ki} T(e_k) = \sum_{k,j} p_{ki} t_{jk} e_j \\
&= \sum_{j,k} t_{jk} p_{ki} e_j,
\end{aligned}
$$

and

$$
\begin{aligned}
L_{e_i} T(e_i) &= e_i \sum_k t_{ki} e_k \\
&= t_{ii} e_i^2 = t_{ii} \sum_k p_{ki} e_k,
\end{aligned}
$$

comparing them, we can have

$$
t_{ii} p_{ji} = \sum_k t_{jk} p_{ki}, \text{ for } j \in \Lambda
$$

$$
t_{ii} p_{ji} = t_{jj} p_{ji}, \text{ for } j \in \Lambda.
$$

Thus, we must have $t_{ii} = t_{jj}$. Therefore,

$$
T(e_i) = k(T) e_i,
$$

where $k(T)$ is a scalar in the ground field K. That is T is a scalar multiplication. So, we can conclude that $\Gamma(E) \cong K$, E is centroidal.

3.3 Nonassociative Banach Algebras

To describe the evolution flow quantitatively in an evolution algebra, it is necessary to introduce a norm. As we will see, under this norm, an evolution algebra becomes a Banach algebra. We will define a norm for an evolution algebra first and then prove that any finite dimensional evolution algebra is a Banach algebra.

3.3.1 Definition of a norm over an evolution algebra

Let E be an evolution algebra with a generator set $\{e_i \mid i \in \Lambda\}$. Define a function N from E to the underlying field K as follows,

$$N : E \longrightarrow K$$

$$N(x) = \sum_i |a_i|,$$

where $x \in E$ and $x = \sum_i a_i e_i$. We can verify that N is a norm as follows:

- Nonnegativity

$$N(x) = \sum_i |a_i| \geq 0.$$

  Furthermore, if $N(x) = 0$, then $N(x) = \sum_i |a_i| = 0$. Thus $|a_i| = 0$, which means that a_i must be 0. That is, $x = 0$. Therefore $N(x) = 0$ if and only if $x = 0$.
- Linearity $N(ax) = |a|N(x), a \in K$, since $N(ax) = \sum_i |aa_i| = |a| \sum_i |a_i| = |a|N(x)$.
- Triangle inequality $N(x + y) \leq N(x) + N(y)$. For $x = \sum_i a_i e_i$ and $y = \sum_i b_i e_i$, we have

$$N(x + y) = N(\sum_i (a_i + b_i)e_i)$$

$$= \sum_i |a_i + b_i|$$

$$\leq \sum_i (|a_i| + |b_i|)$$

$$= \sum_i |a_i| + \sum_i |b_i|$$

$$= N(x) + N(y).$$

Thus, an evolution algebra is a normed algebra. We denote $N(x) = \|x\|$.

Proposition 6. *Any evolution operator L is a bounded linear operator.*

Proof. For $x \in E$, $x = \sum_i a_i e_i$ under a natural basis $\{e_i \mid i \in \Lambda\}$ of an evolution algebra E, we have

$$L(x) = \sum_i a_i L(e_i) = \sum_i a_i e_i^2 = \sum_{i\,j} a_i p_{ji} e_j.$$

$$N(L(x)) = \sum_j |\sum_i a_i p_{ji}| \leq \sum_j \sum_i |a_i p_{ji}|$$

$$\leq \sum_i |a_i| \sum_j |p_{ji}| \leq \sum_i |a_i| c_i$$

$$\leq cN(x),$$

where $c_i = \sum_j |p_{ji}|$, and $c = \max\{c_i \mid i \in \Lambda\}$. Therefore, T is bounded.

Corollary 6. *Each element of $SP(L, E)$ is a bounded linear operator, where $SP(L, E)$ is the linear space of all the operators of left multiplication of E.*

Proof. We know $SP(L, E) = Span(L_i : i \in \Lambda)$ over K. We have $L_i(x) = a_i\, e_i^2$, if $x = \sum_i a_i\, e_i$. Then we can see

$$N(L_i(x)) = N(a_i \sum_j p_{ji}e_j)$$

$$\leq c_i|a_i| \leq c\sum_i |a_i| = cN(x),$$

so L_i is bounded.

Now, $\forall\, \theta \in Sp(L, E)$, write $\theta = \sum_i \beta_i\, L_i$, $\beta_i \in K$. For any $x = \sum a_i\, e_i$, we have

$$\theta(x) = \sum_i \beta_i\, L_i(x) = \sum_i \beta_i\, a_i e_i^2 = \sum_{i\,j} \beta_i\, a_i p_{ji} e_j,$$

then

$$N(\theta(x)) = \sum_{i\,j} |\beta_i\, a_i p_{ji}| \leq c\sum_i |\beta_i\, a_i|$$

$$\leq c\sum_i |\beta_i\,| \cdot \sum_i |a_i|$$

$$\leq cbN(x),$$

where $b = \sum_i |\beta_i\,|$ is a constant for a given operator θ. Therefore θ is bounded.

3.3.2 An evolution algebra as a Banach space

In Functional Analysis, there is a theorem that a linear operator is bounded if and only if it is a continuous operator. From Proposition 6 and Corollary 6, evolution operators and left multiplication operators are all bounded. Therefore, they are continuous under the topology induced by the metric $\rho(x, y) = N(x - y)$, for $x, y \in E$.

Theorem 5. *Let E be an evolution algebra with finite dimension n, then it is complete as a normed linear space. That is, E is a Banach space.*

Proof. Let $x^m = \sum_{i=1}^{n} a_i^m e_i$, $m = 1, 2, \cdots$, be a sequence in E, then we have

$$\rho(a_i^m e_i, a_i^k e_i) = N(a_i^m e_i - a_i^k e_i)$$

$$= |a_i^m - a_i^k| \leq \sum_{i=1}^{n} |a_i^m - a_i^k|$$

$$= \rho(x^m, x^k) \leq n \cdot \max_{1\leq i\leq n} |a_i^m - a_i^k|.$$

When x^m is a Cauchy sequence, then, for any $\varepsilon > 0$, there is an integer m_0, and for any integers m, $k > m_0$, we have $\rho(x^m, x^k) < \varepsilon$. So, we have $|a_i^m - a_i^k| < \varepsilon/n$. By the Cauchy principle in Real Analysis, there is a number b_i, such that $|a_i^m - b_i| < \varepsilon/n$. That is, the coordinate sequence a_i^m converges to b_i, $i = 1, 2, \cdots, n$. If we denote $x^0 = \sum_{i=1}^{n} b_i e_i$, then

$$\rho(x^m, x^0) = \sum_{i=1}^{n} |a_i^m - a_i^k| \leq \varepsilon.$$

This means that x^m converges to x^0. Therefore, E is a complete normed linear space, i.e. E is a Banach space.

Corollary 7. *For a finite dimensional evolution algebra E, it is a nonassociative Banach algebra.*

Proof. It is an immediate consequence of Theorem 5.

Theorem 6. *Let E be a finite dimensional evolution algebra, and $BL(E \to E)$ be the set of all bounded linear operators over E, then the subspace $L(E)$ of $BL(E \to E)$, all left multiplication operators of E, is a Banach subalgebra of $BL(E \to E)$.*

Proof. In Functional Analysis, there is a theorem that when X is Banach space, $\Re(X \longrightarrow X)$, the space of all bounded linear operators from X to X, is a Banach algebra. Because E is a Banach algebra, $BL(E \to E)$ is also a Banach algebra. Since each element of $L(E)$ is bounded and the composite of two elements of $L(E)$ is also bounded, then the operator algebra of left multiplication is a subalgebra of $BL(E \to E)$,

But we know, generally, $L(E)$ is not a Banach subalgebra of $BL(E \to E)$.

3.4 Periodicity and Algebraic Persistency

In this section, we introduce a periodicity for each generator of an evolution algebra. It turns out all generators of a nonnegative simple evolution algebra have the same periodicity. We also introduce an algebraic persistency and an algebraic transiency for each generator of an evolution algebra. They are basic concepts in the study of evolution in algebras.

3.4.1 Periodicity of a generator in an evolution algebra

Definition 6. *Let e_j be a generator of an evolution algebra E, the period d of e_j is defined to be the greatest common divisor of the set $\left\{ \log_2 m \mid e_j < \left(e_j^{(m)} \right) \right\}$, where power $e_j^{(m)}$ is some kth plenary power, $2^k = m$. That is*

$$d = g.c.d. \left\{ \log_2 m \mid e_j < \left(e_j^{(m)} \right) \right\}.$$

If d is 1, we say e_j is aperiodic; if the set $\left\{ \log_2 m \mid e_j < (e_j^{(m)}) \right\}$ is empty, we define $d = \infty$.

To understand this definition, we give a proposition that states relations between evolution operators and plenary powers of an element.

Proposition 7. *Generator e_j has the period d if and only if d is the greatest common divisor of the set $\{n \mid \rho_i \, L^n(e_i) \neq 0\}$. That is*

$$d = g.c.d.\{n \mid \rho_i \, L^n(e_i) \neq 0\},$$

where ρ_i is a projection map of E, which maps every element of E to its e_i component.

Proof. We introduce a notion – plenary powers of a matrix. Let

$$(e_1, e_2, \cdots\cdots, e_n) \cdot (e_1, e_2, \cdots\cdots, e_n)$$
$$= (e_1^2, e_2^2, \cdots\cdots, e_n^2) = (e_1, e_2, \cdots\cdots, e_n)B,$$

where $B = (p_{ij})$ is the structural constant matrix of E.

Look at

$$(e_1^2, e_2^2, \cdots\cdots, e_n^2) \cdot (e_1^2, e_2^2, \cdots\cdots, e_n^2)$$
$$= (e_1, e_2, \cdots\cdots, e_n)B \cdot (e_1, e_2, \cdots\cdots, e_n)B$$
$$= (e_1^{(4)}, e_2^{(4)}, \cdots\cdots, e_n^{(4)})$$
$$= (\sum_k p_{k1}e_k, \ \sum_k p_{k2}e_k, \ \cdots\cdots, \sum_k p_{kn}e_k)$$
$$\cdot(\sum_k p_{k1}e_k, \ \sum_k p_{k2}e_k, \ \cdots\cdots, \sum_k p_{kn}e_k)$$
$$= (\sum_k p_{k1}^2 e_k^2, \ \sum_k p_{k2}^2 e_k^2, \ \cdots\cdots, \sum_k p_{kn}^2 e_k^2)$$
$$= (e_1^2, e_2^2, \cdots\cdots, e_n^2) \begin{pmatrix} p_{11}^2 & p_{21}^2 & \vdots & p_{n1}^2 \\ p_{12}^2 & p_{22}^2 & \vdots & p_{n2}^2 \\ \vdots & \vdots & \vdots & \vdots \\ p_{1n}^2 & p_{2n}^2 & \vdots & p_{nn}^2 \end{pmatrix}$$
$$= (e_1, e_2, \cdots\cdots, e_n)BB^{(2)}.$$

We also compute

$$
\begin{aligned}
&(e_1^{(4)}, e_2^{(4)}, \cdots\cdots, e_n^{(4)}) \\
&\cdot (e_1^{(4)}, e_2^{(4)}, \cdots\cdots, e_n^{(4)}) \\
&= (e_1^{(8)}, e_2^{(8)}, \cdots\cdots, e_n^{(8)}) \\
&= (e_1, e_2, \cdots\cdots, e_n) B B^{(2)} \\
&\quad \cdot (e_1, e_2, \cdots\cdots, e_n) B B^{(2)} \\
&= (e_1, e_2, \cdots\cdots, e_n) B (B B^{(2)})^{(2)}.
\end{aligned}
$$

Now, we define plenary powers for a matrix as follows:

$$
\begin{aligned}
A^{[1]} &= A \\
A^{[2]} &= A A^{(2)} = A (A^{[1]})^{(2)} \\
A^{[3]} &= A (A^{[2]})^{(2)} = A (A A^{(2)})^{(2)}
\end{aligned}
$$

$$
\cdots\cdots\cdots\cdots
$$

$$
A^{[k+1]} = A (A^{[k]})^{(2)}.
$$

Thus, we have

$$
\begin{aligned}
&(e_1^{[m]}, e_2^{[m]}, \cdots\cdots, e_n^{[m]}) \\
&= (e_1, e_2, \cdots\cdots, e_n) B^{[m]}.
\end{aligned}
$$

We note that the matrix representation of the evolution operator L is given by the matrix B. $\rho_j L^k(e_j) \neq 0$ means the (j, j) entry of B^k is not zero. It is not too hard to check that the (j, j) entry of B^k is not zero if and only if the (j, j) entry of $B^{[k]}$ is not zero, which means $\rho_j \left(e_j^{(2^k)} \right) \neq 0$. That also means $e_j < e_j^{(2^k)}$. This concludes the proof.

From the above proof, we can see that the kth plenary power and the kth action of the evolution operator give us the same information in computing the period of an element. We also obtain the following corollary.

Corollary 8. *Generator e_j has the period of d if and only if d is the greatest common divisor of the set $\{ n \mid e_j < e_j^{[n]} \}$, where $e_j^{[n]} = e_j^{(2^n)}$.*

Theorem 7. *All generators have the same period in a nonnegative simple evolution algebra.*

Proof. Let e_i and e_j be two generators in a simple evolution algebra E. The periods of e_i and e_j are d_i and d_j respectively. Since e_i must occur in a plenary power of e_j, say $e_i < e_j^{[n]}$, and e_j must occur in a plenary power of e_i, say $e_j < e_i^{[m]}$, from Theorem 3 we have $e_i < e_i^{[n+m]}$ and $e_j < e_j^{[n+m]}$. Then $d_i \mid n+m$, and $d_j \mid n+m$. Since $e_j < e_j^{[d_j]}$, so $e_i < e_j^{[d_j+n]}$ and $e_i < e_i^{[d_j+n+m]}$, then $d_i \mid d_j + n + m$. Therefore $d_i \mid d_j$. Similarly, we have $d_j \mid d_i$. Thus, we get $d_i = d_j$.

3.4.2 Algebraic persistency and algebraic transiency

Let E be an evolution algebra with a generator set $\{e_i \mid i \in \Lambda\}$. We say that generator e_j is *algebraically persistent* if the evolution subalgebra $\langle e_j \rangle$, generated by e_j, is a simple subalgebra, and e_i is *algebraically transient* if the subalgebra $\langle e_i \rangle$ is not simple. Then, it is obvious that every generator in a simple evolution algebra is algebraically persistent, since each generator generates the same algebra that is simple. We know that if x and y intercommunicate, the evolution subalgebra generated by x is the same as the one generated by y. Moreover, we have the following theorem.

Theorem 8. *Let e_i and e_j be generators of an evolution algebra E. If e_i and e_j can intercommunicate and both are algebraically persistent, then they belong to the same simple evolution subalgebra of E.*

Proof. Since e_i and e_j can intercommunicate, e_i occurs in $\langle e_j \rangle$ and e_j occurs in $\langle e_i \rangle$. Then, there are some powers of e_i, denoted by $P(e_i)$ and some powers of e_j, denoted by $Q(e_j)$, such that

$$P(e_i) = ae_j + u \qquad a \neq 0,$$
$$Q(e_j) = be_i + v \qquad b \neq 0.$$

Since subalgebras are also ideals in an evolution algebra, we have

$$P(e_i)e_j = ae_j^2 \in \langle e_i \rangle,$$
$$Q(e_j)e_i = ae_i^2 \in \langle e_j \rangle.$$

Therefore, $\langle e_i \rangle \cap \langle e_j \rangle \neq \{0\}$. Since $\langle e_i \rangle$ and $\langle e_j \rangle$ are both simple evolution subalgebras, then $\langle e_i \rangle = \langle e_j \rangle$. Thus, e_i and e_j belong to the same simple evolution subalgebra. $\qquad \blacksquare$

For an evolution algebra, we can give certain conditions to specify whether it is simple or not by the following corollary:

Corollary 9. *1) Let E be a connected evolution algebra, then E has a proper evolution subalgebra if and only if E has an algebraically transient generator.*

2) Let E be a connected evolution algebra, then E is a simple evolution algebra if and only if E has no algebraically transient generator.

3) If E has no algebraically transient generator, then E can be written as a direct sum of evolution subalgebras (the number of summands can be one).

Proof. 1) If E has no algebraically transient generator, each generator e_i generates a simple evolution subalgebra. These subalgebras are all the same because E is connected. Otherwise, E would be a direct sum of these subalgebras. This means the only nonempty subalgebra of E is itself. On the other hand, if E has an algebraically transient generator e_k, then the generated evolution subalgebra $\langle e_k \rangle$ is not simple. This means $\langle e_k \rangle$ has a proper subalgebra, so E has a proper subalgebra.

2) It is obvious from (1).

3) It is also obvious from (1).

Now, the question is, for any evolution algebra, whether there is always an algebraically persistent generator. Generally, this is not true. The following statement tells us that for any finite dimensional evolution algebra, there always is an algebraically persistent generator.

Theorem 9. *Any finite dimensional evolution algebra has a simple evolution subalgebra.*

Proof. We assume the evolution algebra E is connected, otherwise we just need to consider a component of a direct sum of E.

Let $\{e_1, e_2, \cdots, e_n\}$ be a generator set of E. Consider evolution subalgebras generated by each generator

$$\langle e_1 \rangle, \quad \langle e_2 \rangle, \quad \cdots \cdots, \langle e_n \rangle.$$

If there is a subalgebra that is simple, it is done. Otherwise, we choose a subalgebra that contains the least number of generators, for example, $\langle e_i \rangle$ and $\{e_{i_1}, e_{i_2}, \cdots, e_{i_k}\} \subset \langle e_i \rangle$, where $\{e_{i_1}, e_{i_2}, \cdots, e_{i_k}\}$ is a subset of $\{e_1, e_2, \cdots, e_n\}$. Then, consider

$$\langle e_{i_1} \rangle, \quad \langle e_{i_2} \rangle, \quad \cdots \cdots, \langle e_{i_k} \rangle.$$

If there is some subalgebra that is simple in this sequence, we are done. Otherwise, we choose a certain $\langle e_{i_j} \rangle$ in the same way as we choose $\langle e_i \rangle$. Since the number of generators is finite, this process will stop. Therefore, we always have a simple evolution subalgebra. Of course, any generator of the simple evolution subalgebra is algebraically persistent.

3.5 Hierarchy of an Evolution Algebra

The hierarchical structure of an evolution algebra is a remarkable property that gives a picture of the dynamical process when multiplication in the evolution algebra is treated as a discrete time dynamical step. In this section, we study this hierarchy and establish a principal theorem about evolution algebras – the hierarchical structure theorem. Algebraically, this hierarchy is a sequence of semidirect-sum decompositions of a general evolution algebra. It depends upon the "relative" concepts of algebraic persistency and algebraic transiency. By the "relative" concepts here, we mean that we can define higher algebraic persistency and algebraic transiency over the space generated by transient generators in the previous level. The difference between algebraic persistency and algebraic transiency suggests a sequential semidirect-sum decomposition, or suggests a direction of evolution from the viewpoint of dynamical systems. This hierarchical structure demonstrates that our evolution algebra is a mixed algebraic and dynamical subject. We also establish the structure theorem for simple evolution algebras. A method is given here to reduce a "big" evolution algebra to a "smaller" one, with the hierarchy being the same. This procedure is called reducibility, which gives a rough classification of all evolution algebras – the skeleton-shape classification.

3.5.1 Periodicity of a simple evolution algebra

As we know in Section 3.4 Theorem 7, all generators of a nonnegative simple evolution algebra have the same period. It might be well to say that a simple algebra has a period. Thus, simple evolution algebras can be roughly classified as either periodic or aperiodic. The following theorem establishes the structure of a periodic simple evolution algebra.

Theorem 10. *Let E be a nonnegative simple evolution algebra with generator set $\{e_i \mid i \in \Lambda\}$, then all generators have the same period, denoted by d. There is a partition of generators with d disjointed classes C_0, C_2, $\cdots$, C_{d-1}, such that $L(\Delta_k) \subseteq \Delta_{k+1}(modd)$, or $\Delta_k^2 \subseteq \Delta_{k+1}(modd)$, $k = 1, 2, \cdots d - 1$, where $\Delta_k = Span\,(C_k)$ and L is the evolution operator of E, mod is taken with respect to the index of the class of generators. There is also a direct sum of linear subspaces*

$$E = \Delta_0 \oplus \Delta_1 \oplus \cdots \oplus \Delta_{d-1}.$$

Proof. Since E is simple, if any generator e_i has a period of d, then every generator has a period of d. Set $C_m = \left\{ e_j \mid e_j < e_i^{[nd+m]}, j \in \Lambda \right\}, 0 \le m < d,$ for any fixed e_i. Because this evolution algebra is simple, each generator e_j will occur in some C_m. So

$$\cup_{m=0}^{d-1} C_m = \{e_k \mid k \in \Lambda\}.$$

Claim that these C_m are disjoint. We show this as follows: if $e_j \in C_{m_1} \cap C_{m_2}$ for $0 \le m_1, m_2 < d$, then $e_j < e_i^{[n_1 d + m_1]}$, and $e_j < e_i^{[n_2 d + m_2]}$ for some integers n_1 and n_2. Since $\langle e_i \rangle = \langle e_j \rangle$, so $e_i < \langle e_j \rangle$. That is, $e_i < e_j^k$ for some integer k. Therefore $e_i < e_i^{[n_1 d + m_1 + k]}$, and $e_i < e_i^{[n_2 d + m_2 + k]}$, then we have $d \mid n_1 d + m_1 + k$, and $d \mid n_2 d + m_2 + k$. Thus $d \mid m_1 - m_2$. But $0 \le |m_1 - m_2| < d$, so we have $m_1 = m_1$, then $C_{m_1} = C_{m_2}$.

Therefore, a partition of the set $\{e_k \mid k \in \Lambda\}$ is obtained. We need to prove that if we take e_k as a fixed generator that is different from the previous e_i for partitioning, we can still get the same partition. Fix e_k, let $C'_m = \left\{ e_j \mid e_j < e_k^{[nd+m]}, j \in \Lambda \right\}$, where $0 \le m < d$. Since E is simple, $e_i < e_k^{[t]}$. If $e_\alpha, e_\beta \in C_m$, then $e_\alpha < e_i^{[n_1 d + m]}$, and $e_\beta < e_i^{[n_2 d + m]}$ for some integers n_1 and

n_2. Then $e_\alpha < e_k^{[n_1 d + m + t]}$, $e_\beta < e_k^{[n_2 d + m_2 + k]}$. Since $n_1 d + m + t \equiv n_2 d + m + t$ (*modd*), so e_α and e_β are still in the same cell C'_m of the partition.

Now, if $e_j \in C_k$, then $e_i^{(2^{nd+k})} = a e_j + v$, $a \neq 0$. We have $e_i^{[k+1]} = a^2 e_j^2 + v^2 = a^2 L(e_j) + v^2$, which means that generators occur in $L(e_j) \in C_{k+1}$ or generators occur in $e_j^2 \in C_{k+1}$.

Denote the linear subspace spanned by C_k as Δ_k, $k = 0, 1, 2, \cdots d - 1$, then we have a direct sum for E

$$E = \Delta_0 \oplus \Delta_1 \oplus \cdots \oplus \Delta_{d-1},$$

and

$$L: \quad \Delta_k \to \Delta_{k+1} \qquad k = 1, 2, \cdots d - 1;$$
$$L^d: \quad \Delta_k \to \Delta_k, \text{ a linear map for each } k.$$

Or, we have

$$\Delta_k^2 \subseteq \Delta_{k+1}, \qquad \Delta_k^d \subseteq \Delta_k, \qquad k = 1, 2, \cdots d - 1.$$

This concludes the proof.

3.5.2 Semidirect-sum decomposition of an evolution algebra

A general evolution algebra has algebraically persistent generators and algebraically transient generators. These two types of generators have distinct "reproductive behavior" – dynamical behavior. Algebraically persistent ones can generate a simple subalgebra. Once an element belongs to the subalgebra, it will never "reproduce" any element that is not in the subalgebra. Or, dynamically, once the dynamical process, represented by the evolution operator L, enters a simple evolution subalgebra, it will never escape from it. In contrast, algebraically transient generators behave differently. They generate reducible subalgebras. The following theorem demonstrates how to distinguish these two types of generators algebraically. Actually, it is the starting level of the hierarchy of an evolution algebra, and it can also serve as a sample of structure in each level.

Theorem 11. *Let E be a connected evolution algebra. As a vector space, E has a decomposition of direct sum of subspaces:*

$$E = A_1 \oplus A_2 \oplus \cdots \oplus A_n \overset{\bullet}{+} B,$$

where A_i, $i = 1, 2, \cdots, n$, are all simple evolution subalgebras, $A_i \cap A_j = \{0\}$ for $i \neq j$, and B is a subspace spanned by algebraically transient generators (which we call a transient space). The summation $A_1 \oplus A_2 \oplus \cdots \oplus A_n$ is a direct sum of subalgebras. Symbol $\overset{\bullet}{+}$ indicates the summation is not a direct sum of subalgebras, just a direct sum of subspaces. We call this decomposition a semidirect-sum decomposition of an evolution algebra.

Proof. Take a generator set for E, $\{e_i \mid i \in \Lambda\}$, where Λ is a finite index set, then we will have two categories of generators: algebraically transient generators and algebraically persistent generators. Let

$$B = Span\,(e_k \mid e_k \text{ is algebraically transient})\,.$$

Take any algebraically persistent element e_{i_1}, let $A_1 = \langle e_{i_1} \rangle$. Again take any algebraically persistent element e_{i_2} that does not occur in A_1, let $A_2 = \langle e_{i_2} \rangle$. Keep doing in this way. Since Λ is finite, we will end up with some $A_n = \langle e_{i_n} \rangle$.

By our construction, each A_k is simple, since e_{i_k} is algebraically persistent. And $A_i \cap A_j = \{0\}$ for $i \neq j$, since they are simple. Finally, as a vector space E, $A_1 \oplus A_2 \oplus \cdots \oplus A_n \overset{\bullet}{+} B$ is a direct sum decomposition, since $A_i \cap B = \{0\}, i = 1, 2, \cdots, n$. But B is not a subalgebra; it is just a linear subspace. Therefore, as an algebra E, we just say that it is a semidirect-sum decomposition.

Note, if E is simple, n is 1 and $B = \phi$. Otherwise, B is not zero.

3.5.3 Hierarchy of an evolution algebra

1). The *0th* structure of an evolution algebra E : the *0th* decomposition of E is given by Theorem 11 as

$$E = A_1 \oplus A_2 \oplus \cdots \oplus A_{n_0} \overset{\bullet}{+} B_0,$$

where B_0 is the subspace spanned by algebraically transient generators of E, we call it the *0th* transient space.

2). The *1st* structure of E, which is the decomposition of the *0th* transient space B_0.

Although the *0th* transient space B_0 is not an evolution subalgebra, it inherits evolution algebraic structure from E if the algebraic multiplication is confined within B_0. We shall make this point clear.

- The induced multiplication: we write generators for B_0 as $e_{0,k}$ and $k \in \Lambda_0$, where $\Lambda_0 \subset \Lambda$ is a subset of the index set. Actually, they are algebraic transient generators. Then, we have the induced multiplication on B_0, denoted by $\overset{1}{\cdot}$, as follows

$$e_{0,i} \overset{1}{\cdot} e_{0,j} = 0 \quad \text{if } i \neq j,$$

$$e_{0,i} \overset{1}{\cdot} e_{0,i} = \rho_{B_0}(e_{0,i} \cdot e_{0,i}),$$

and linearly extend onto $B_0 \times B_0$, where ρ_{B_0} is the projection from E to B_0. It is not hard to check that B_0 is an evolution algebra, which we call the first induced evolution algebra.

- The first induced evolution operator in B_0 is given by

$$L_{B_0} = \rho_{B_0} L.$$

Then, we have

$$L_{B_0}^2 = (\rho_{B_0} L)(\rho_{B_0} L) = \rho_{B_0} L^2,$$

and for any positive integer n, we have

$$L_{B_0}^n = \rho_{B_0} L^n.$$

- First induced evolution subalgebras generated by some generators of B_0:
Denote the evolution subalgebra generated by $e_{0,i}$ in B_0 by $\langle e_{0,i} \mid B_0 \rangle$ (using multiplication $\overset{1}{\cdot}$ in B_0). Sometimes we just use $\langle e_{0,i} \rangle_1$ for this subalgebra. (It may be a nilpotent subalgebra).
- First algebraically persistent generators in B_0:
We say $e_{0,i}$ is a first algebraically persistent if $\langle e_{0,i} \rangle_1$ is a simple subalgebra. Otherwise, we say $e_{0,i}$ is a first algebraically transient.
B_0 is called irreducible (simple) if it has no proper first induced evolution subalgebra. Similarly, we have a first reducible evolution subalgebra.
B_0 is connected if B_0 can not be decomposed as a direct sum of two first induced evolution subalgebras.

- The *1st* decomposition of E, the decomposition of B_0:
We state the decomposition theorem for the *0th* transition space B_0 here. The proof is essentially a repeat of that of the *0th* decomposition theorem. We therefore skip the proof.

Theorem 12. *The* 1st *structure of an evolution algebra* E : *the* 1st *decomposition of* E *is given by*

$$B_0 = A_{1,1} \oplus A_{1,2} \oplus A_{1,3} \oplus \cdots \oplus A_{1,n_1} \overset{\bullet}{+} B_1$$

where $A_{1,i}$, $i = 1, 2, \cdots, n_1$, *are all first simple evolution subalgebras of* B_0, $A_{1,i} \cap A_{1,j} = \{0\}$, *if* $i \neq j$, *and* B_1 *is the first transient space spanned by the first algebraically transient generators.*

- The first induced periodicity and intercommunication:
The following is the definition of the first induced period

$$\text{The period of } e_{0,i} = \gcd\{n \mid e_{0,i} < e_{0,i}^{[n]_o}\},$$

where $e_{0,i}^{[n]_o}$ means that the plenary powers are taken within space B_0. We have a theorem about the intercommunications within the space B_0. The proof is the same as that at the *0th* level. We will not give it here.

Theorem 13. *If $e_{0,i}$ and $e_{0,j}$ intercommunicate, then they have the same first induced periods.*

- The decomposition of a first simple periodical evolution subalgebra:

Theorem 14. *If $A_{1,k}$ is a first nonnegative simple periodic reduced evolution subalgebra and some $e_{0,i}$ of its generator has a period of d, then it can be written as a direct sum*

$$A_{1,k} = \Delta_{1,0} \oplus \Delta_{1,1} \oplus \cdots \oplus \Delta_{1,d-1}.$$

(The proof is the same as that in the *0th* level.)

3). We can construct the *2nd* induced evolution algebra over the first transient space B_1, if B_1 is connected and not simple. If the *kth* transient space B_k is disconnected and each component is simple, we will stop with a direct sum of *(k+ 1)th* simple evolution subalgebras. Otherwise, we can continue to construct evolution subalgebras until we reach a level where each evolution subalgebra is simple. Now, we have the hierarchy as follows

$$E = A_{0,1} \oplus A_{0,2} \oplus \cdots \oplus A_{0,n_0} \overset{\bullet}{+} B_0$$
$$B_0 = A_{1,1} \oplus A_{1,2} \oplus \cdots \oplus A_{1,n_1} \overset{\bullet}{+} B_1$$
$$B_1 = A_{2,1} \oplus A_{2,2} \oplus \cdots \oplus A_{2,n_2} \overset{\bullet}{+} B_2$$
$$\cdots\cdots\cdots\cdots\cdots\cdots\cdots\cdots\cdots$$
$$B_{m-1} = A_{m,1} \oplus A_{m,2} \oplus \cdots \oplus A_{m,n_m} \overset{\bullet}{+} B_m$$
$$B_m = B_{m,1} \oplus B_{m,2} \oplus \cdots \oplus B_{m,h},$$

where $A_{k,l}$ is a *kth* simple evolution subalgebra, $A_{k,l} \cap A_{k,l\prime} = \{0\}$ if $l \neq l\prime$, B_k is the *kth* transient space. B_m can be decomposed as a direct sum of *(m+ 1)th* simple evolution subalgebras. We may call these *(m+ 1)th* simple evolution subalgebras the heads of the hierarchy, and h is the number of heads.

Example 2. Let's look at an evolution algebra E with dimension 25. The generator set is $e_1, e_2, \cdots, e_{25}$. The defining relation are given: $e_i e_j = 0$ if $i \neq j$; when $i = j$, they are

$$e_1^2 = e_2 + 2e_3 + e_4 + 3e_5, \quad e_2^2 = 2e_3 + 7e_6 + e_9,$$
$$e_3^2 = e_2 + 5e_7 + e_8 + 9e_9, \quad e_4^2 = 7e_5 + e_9 + e_{10} + 10e_{11},$$
$$e_5^2 = e_4 + 7e_9 + 5e_{12}, \quad e_6^2 = e_7 + e_8 + 7e_{13},$$
$$e_7^2 = 6e_6 + e_8 + 2e_{13}, \quad e_8^2 = e_6 + 3e_7 + e_{13} + 2e_{14},$$
$$e_9^2 = 3e_{15} + 2e_{14}, \quad e_{10}^2 = 4e_{11} + e_{12} + 2e_{16},$$
$$e_{11}^2 = 6e_{10} + e_{12} + 5e_{15}, \quad e_{12}^2 = e_{10} + 4e_{11} + 2e_{15} + e_{16},$$
$$e_{13}^2 = e_{14} + 5e_{17} + 3e_{18} + e_{21},$$

$$e_{14}^2 = e_{13} + 4e_{17} + e_{18} + 5e_{19} + e_{20},$$
$$e_{15}^2 = 8e_{16} + e_{20} + e_{21} + 7e_{22},$$
$$e_{16}^2 = 9e_{15} + e_{23} + 10e_{24} + e_{25},$$
$$e_{17}^2 = 3e_{17} + 2e_{18}, \ e_{18}^2 = 4e_{17} + 2e_{18}, \ e_{19}^2 = 3e_{19} + e_{20},$$
$$e_{20}^2 = e_{19}, \ e_{21}^2 = 3e_{22} + e_{21}, \ e_{22}^2 = 2e_{22} + 5e_{21},$$
$$e_{23}^2 = e_{25} + 4e_{24}, \ e_{24}^2 = 2e_{25}, \ e_{25}^2 = e_{23} + 8e_{24}.$$

The 0th evolution subalgebras are $A_{0,1} = \langle e_{17}, e_{18} \rangle$, $A_{0,2} = \langle e_{19}, e_{20} \rangle$, $A_{0,3} = \langle e_{21}, e_{22} \rangle$, and $A_{0,4} = \langle e_{23}, e_{24}, e_{25} \rangle$. The 0th transient space is $span e_1, e_2, \cdots, e_{16}$. The 1st evolution subalgebras are $A_{1,1} = \langle e_{13}, e_{14} \rangle$ and $A_{1,2} = \langle e_{15}, e_{16} \rangle$. The 2nd evolution subalgebras are $A_{2,1} = \langle e_6, e_7, e_8 \rangle$, $A_{2,2} = \langle e_9 \rangle$, and $A_{2,3} = \langle e_{10}, e_{11}, e_{12} \rangle$. The 3rd evolution subalgebras are $A_{3,1} = \langle e_2, e_3 \rangle$ and $A_{3,2} = \langle e_4, e_5 \rangle$. The 3rd transient space, the head of the hierarchy given by the algebra B_3, is $span\{e_1\}$. Figure 3.1 shows the hierarchical structure.

3.5.4 Reducibility of an evolution algebra

From the hierarchy of an evolution algebra, we get an impression about the dynamical flow of an algebra. That is, if we start at a high level, a big index level, the dynamical flow will automatically go down to a low level, it may also sojourn in a simple evolution subalgebra at each level. It is reasonable to view each simple evolution subalgebra at each level as one point or one-dimensional subalgebra. The big evolution picture still remains. If we call this remained hierarchy the *skeleton* of the original evolution algebra, all evolution algebras that possess the same skeleton will have a similar dynamical behavior. We call this procedure the *reducibility* of an evolution algebra and write it as a statement.

Theorem 15. *Every evolution algebra E can be reduced to a unique evolution algebra E_r such that its evolution subalgebras in its hierarchy are all one-dimensional subalgebras. We call such a unique evolution algebra E_r the skeleton of E.*

Example 3. The skeleton E_r of the algebra E in Example 2 is the evolution algebra generated by $\eta_1, \eta_2, \cdots, \eta_{12}$ that are subject to the following defining relations:

$$\eta_1^2 = \eta_2 + \eta_3, \ \eta_2^2 = \eta_4 + \eta_5, \ \eta_3^2 = \eta_5 + \eta_6,$$
$$\eta_4^2 = \eta_7, \ \eta_5^2 = \eta_7 + \eta_8, \ \eta_6^2 = \eta_8,$$
$$\eta_7^2 = \eta_9 + \eta_{10} + \eta_{11}, \ \eta_8^2 = \eta_{12} + \eta_{10} + \eta_{11},$$
$$\eta_9^2 = \eta_9, \ \eta_{10}^2 = \eta_{10}, \ \eta_{11}^2 = \eta_{11}, \ \eta_{12}^2 = \eta_{12}.$$

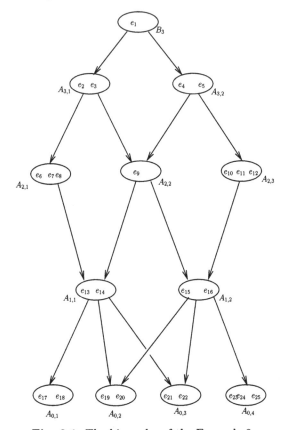

Fig. 3.1. The hierarchy of the Example 2

The Fig. 3.2 shows the hierarchical structure of E_r. Comparing with Fig. 3.1, these two have the same dynamical shape.

The concept of the skeleton can be utilized to give a rough classification of all evolution algebras. From Examples 2 and 3, we can see that two types of numbers, the number of levels m and the numbers n_k of simple evolution subalgebras at each level k, can roughly determine the shape of the hierarchy of an evolution algebra, ignoring the flow relations between two different levels. Note that at level $(m+1)$, the number n_{m+1} is h, the number of heads, in our notation. We give the criterions for *classification of evolution algebras*. That is, if two evolution algebras have the same number m of levels and the numbers n_k of simple evolution subalgebras at each level k, we say these two evolution algebras belong to the same *class of skeleton-shape*. Furthermore, we say two evolution algebras belong to the same *class of skeleton* if they belong to the same class of skeleton-shape and the flow relations between any two different levels are the same correspondingly.

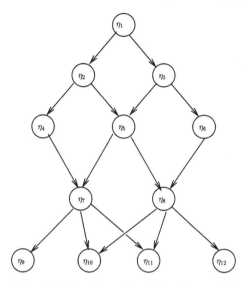

Fig. 3.2. The hierarchy of the Example 3

Now, there are two basic questions related to our classifications that should be answered.

The first one stated as follows: given the level number m and the total number n of simple evolution subalgebras (including heads) wherever they are, how many classes of skeleton-shapes of evolution algebras can we have? The answer is a famous number in number theory, $p_{m+1}(n)$, the number of partitions of n into $m+1$ cells. For $n < m+1$, $p_{m+1}(n) = 0$ and $p_{m+1}(m+1) = 1$. Generally, we have the recursion

$$p_{m+1}(n) = p_{m+1}(n-m-1) + p_m(n-m-1) + \cdots + p_1(n-m-1).$$

We list here the answers for the question when the hierarchy has small levels as follows:

$$p_1(n) = 1, \ m = 0;$$

$$p_2(n) = \begin{cases} \frac{n}{2}, & n \equiv 0\,(2), \\ \frac{n-1}{2}, & n \equiv 1\,(2), \end{cases} \ m = 1;$$

$$p_3(n) = \begin{cases} \frac{n^2}{12}, & n \equiv 0\,(6), \\ \frac{n^2}{12} - \frac{1}{12}, & n \equiv 1\,(6), \\ \frac{n^2}{12} - \frac{1}{3}, & n \equiv 2\,(6), \\ \frac{n^2}{12} + \frac{1}{4}, & n \equiv 3\,(6), \\ \frac{n^2}{12} - \frac{1}{3}, & n \equiv 4\,(6), \\ \frac{n^2}{12} - \frac{1}{12}, & n \equiv (6), \end{cases} \ m = 2.$$

Generally, we have

$$p_{m+1}(n) = \frac{n^m}{m!\,(m-1)!} + R_{m-1}(n), \; n \equiv n' \left((m+1)! \right),$$

where $R_{m-1}(n)$ is a polynomial in n of degree at most $m-1$. Therefore, by the number of levels and the numbers of simple evolution subalgebras, we can determine any evolution algebra up to its skeleton-shape. Thus, we obtain a skeleton-shape classification of all evolution algebras.

The second problem is that, given the level number m and the numbers n_k of simple evolution subalgebras at each level, how many classes of skeletons of evolution algebras can we have? We will use a formula that gives the number $bp(n,m)$ of bipartite graphs with two given disjoint vertex sets, V_1 and V_2, and $|V_1| = n$ $|V_2| = m$. This formula is given by Winfried Just:

$$bp(n,m) = \sum_{k=0}^{n} (-1)^k \binom{n}{k} \left(\sum_{l=0}^{m} (-1)^l \binom{m}{l} 2^{(n-k)(m-l)} \right).$$

Then, the number of classes of skeletons of evolution algebras with m levels and n_k subalgebras at each level is

$$bp(n_0, n_1) bp(n_1, n_2) \cdots bp(n_{m-1}, n_m) = \prod_{i=1}^{m-1} bp(n_i, n_{i+1}).$$

Therefore, by the number of levels and subalgebras at each level, we can determine any evolution algebra up to its skeleton.

4

Evolution Algebras and Markov Chains

For a Markov chain, we can define an evolution algebra by taking states as generators and transition probability vectors as defining relations. We may say an evolution algebra defined by a Markov chain is a Markov evolution algebra. Every property of a Markov chain can be redefined by its Markov evolution algebra. In other words, properties of Markov chains can be revealed by studying their evolution algebras. Moreover, Markov chains, as a type of dynamical systems, have a hidden algebraic aspect. In first three sections of this chapter we study the relations between Markov chains and evolution algebras. In the last section, the hierarchy of a general Markov chain is revealed naturally by its evolution algebra.

4.1 A Markov Chain and Its Evolution Algebra

In this section, let us recall some basic properties of Markov chains and define an evolution algebra for a discrete time Markov chain.

4.1.1 Markov chains (discrete time)

A stochastic process $X = \{X_0, X_1, X_2, \cdots\}$ is a Markov chain if it satisfies Markov property

$$\Pr\{X_n = s_n \mid X_0 = s_0, \ X_1 = s_1, \cdots, \ X_{n-1} = s_{n-1}\}$$
$$= \Pr\{X_n = s_n \mid X_{n-1} = s_{n-1}\}$$

for all $n \geq 1$ and all $s_i \in S$, where $S = \{s_i \mid i \in \Lambda\}$ is a finite or countable infinite set of states. Note that there is an underlying probability space (Ω, ξ, P) for the Markov chain.

The chain X is called homogeneous if

$$
\begin{aligned}
&\Pr\left\{X_n = s_n \mid X_{n-1} = s_{n-1}\right\} \\
&= \Pr\left\{X_{n+k} = s_n \mid X_{n+k-1} = s_{n-1}\right\}
\end{aligned}
$$

for $k = -(n-1), (n-2), \cdots, -1, 0, 1, 2, \cdots$. That is, the transition probabilities $p_{ij} = \Pr\left\{X_{n+1} = s_i \mid X_n = s_j\right\}$ are invariant, i.e., do not depend on n.

4.1.2 The evolution algebra determined by a Markov chain

A Markov chain can be considered as a dynamical system as follows. Suppose that there is a certain mechanism behind a Markov chain, and view this mechanism as a reproductive process. But it is a very special case of reproduction. Each state can be considered as an allele. They just "cross" with itself, and different alleles (states) can not cross or they cross to produce nothing. We introduce a multiplication for the reproduction. Thus we can define an algebraic system that can describe a Markov chain. The multiplication for states is defined to be $e_i \cdot e_i = \sum_k p_{ki} e_k$ and $e_i \cdot e_j = 0$, $(i \neq j)$. It turns out that this system is an evolution algebra. Thus, we have the following theorem.

Theorem 16. *For each homogeneous Markov chain X, there is an evolution algebra M_X whose structural constants are transition probabilities, and whose generator set is the state space of the Markov chain.*

In what follows, we will use the notation M_X for the evolution algebra that corresponds to the Markov chain X. As we see, the constraint for this type of evolution algebra is that

$$
\sum_k p_{ki} = 1, \text{ and}
$$

$$
0 \leq p_{ki} \leq 1.
$$

As we defined in Chapter 3, this type of evolution algebra is called Markov evolution algebra. If we recall the definition of evolution operators in the previous chapter, it is easy to see the following corollary.

Corollary 10. *Let M_X be the evolution algebra corresponding to the Markov chain X with the state set $\{e_i \mid i \in \Lambda\}$ and the transition probability $p_{ij} = \Pr\left\{X_n = e_i \mid X_{n-1} = e_j\right\}$, then the matrix representation of the evolution operator is the transpose of the transition probability matrix.*

Proof. We recall the definition of the evolution operator that $L(e_i) = e_i^2 = \sum_k p_{ki} e_k$, then its matrix representation is given by

$$\begin{pmatrix} p_{11} & p_{12} & \cdot & p_{1n} & \cdot \\ p_{21} & p_{22} & \cdot & p_{2n} & \cdot \\ \vdots & \vdots & \vdots & \vdots & \vdots \\ p_{n1} & p_{n2} & \cdot & p_{nn} & \cdot \\ \vdots & \vdots & \vdots & \vdots & \vdots \end{pmatrix}.$$

The transition probability matrix of the Markov chain is

$$\begin{pmatrix} p_{11} & p_{21} & \cdots & p_{n1} & \cdots \\ p_{12} & p_{22} & \cdots & p_{n2} & \cdots \\ \cdots & \cdots & \cdots & \cdots & \cdots \\ p_{1n} & p_{2n} & \cdots & p_{nn} & \cdots \\ \cdots & \cdots & \cdots & \cdots & \cdots \end{pmatrix}.$$

So the matrix representation of the evolution operator L is a column stochastic matrix.

The evolution operator can be utilized to describe the full range of possible motions of a Markov chain (or, a particle) over its states. It can be viewed as a representation of a dynamical source behind the Markov chain. From this viewpoint, a Markov chain can also be viewed as a linear dynamical system over an algebra. In fact, we can treat a Markov chain as a linear dynamical system L. Thus, we will have a new version of the Chapman–Kolmogorov equation. Before discussing Chapman–Kolmogorov equation, we need a lemma about evolution operators.

Lemma 4. *Let X be a Markov chain if the initial variable X_0 has the mass function v_0, then X_n's mass function v_n can be obtained by the evolution operator of the evolution algebra M_X, $v_n = L^n(v_0)$.*

Proof. The proof depends on the relation between the Markov chain and its evolution operator.

Since the state set is at most countable, the mass function v_0 of X_0 is a vector, which is $v_0 = \sum_i a_i e_i$, where $\{e_i \mid i \in \Lambda\}$ is the state set. It is clear that at any time instant or step, the mass function of X_n is always a vector of this form whose coefficients are all nonnegative and sum to one. Denote v_n as the mass function of X_n. We have $L(v_0) = v_1$, $L^2(v_0) = L(v_1) = v_2$ and so on. This is because

$$L(v_0) = L(\sum_i a_i e_i) = \sum_i a_i L(e_i)$$
$$= \sum_i a_i \sum_k p_{ki} e_k = \sum_{i\,k} a_i p_{ki} e_k$$
$$= \sum_k (\sum_i p_{ki} a_i) e_k;$$

on the other hand, in probability theory

$$\Pr\{X_1 = e_k\}$$
$$= \sum_i \Pr\{X_1 = e_k \mid X_0 = e_i\} \Pr\{X_0 = e_i\}$$
$$= \sum_i p_{ik} a_i.$$

Therefore, we have $L(v_0) = v_1$. Similarly, we can get any general probability vector v_n by the operator L.

As we know, at each epoch n, the position of a Markov chain is described by the possible distribution over the state set $\{e_i \mid i \in \Lambda\}$ (the mass function of X_n). If we view the probability vectors, which are of the form $\sum_i a_i e_i$ subject to $0 \le a_i \le 1$ and $\sum_i a_i = 1$, as general states, we may call the original states "characteristic states" and have the compact cone in the Banach space M_X as the "state space" of the Markov chain. The trace of the Markov chain is a real path in this compact cone.

4.1.3 The Chapman–Kolmogorov equation

Given a Markov chain X, we have a corresponding evolution algebra M_X. For the evolution operator L of M_X, it seems trivial that we have the following formulae of composition of operator L:

$$L^{l+m} = L^l \circ L^m, \tag{4.1}$$

or

$$L^{(r+n+m,\, m)} = L^r \circ L^{(n+m,\, m)}, \tag{4.2}$$

where $L^{(r,\, m)} = L^r \circ L^m$, starting at the mth power, and l, m, n, r are all nonnegative integers. In terms of generators (states), we have

$$\left\| \rho_j \, L^{l+m}(e_i) \right\| = \sum_k \left\| \rho_j \, L^l(e_k) \right\| \cdot \left\| \rho_k \, L^m(e_i) \right\|. \tag{4.3}$$

Remember, our norm in the algebra M_X has a significance of probability. That is, if $v = \sum_i a_i e_i$, then $\|v\|$ can be interpreted as the probability of the vector v presented. The action of the evolution operator can be interpreted as the moving of the Markov chain. Then, the left-hand side of the above equation 4.3 represents the probability of going from e_i to e_j in $l + m$ steps. This amounts to measuring the probability of all these sample paths that start at e_i and end at e_j after $l + m$ steps. The right-hand side takes the collection of paths and partitions it according to where the path is after l steps. All these paths that go from e_i to e_k in l steps and then from e_k to e_j in m steps are grouped together and the probability of this group of paths is given by

$\|\rho_j\, L^l(e_k)\| \cdot \|\rho_k\, L^m(e_i)\|$. By summing these probabilities over all e_k, $k \in \Lambda$, we get the probability of going from e_i to e_j in $l + m$ steps. That is, in going from e_i to e_j in $l+m$ steps, the chain must be in some place in the state space after l steps. The right-hand side of the equation considers all the places it might be in and uses this as a criterion for partitioning the set of paths that are from e_i to e_j in $l+m$ steps. Thus, the above three equations 4.1, 4.2, and 4.3 are all versions of the Chapman–Kolmogorov equation.

We can give a concrete proof about our version of the Chapman–Kolmogorov equation as follows. Since we work on an evolution algebra, it is natural for us to use matrix representation of evolution operators.

Proof. Let the matrix representation of the evolution operator L be $A = (p_{ji})$

$$\rho_j\, L(e_i) = p_{ji} e_j \Rightarrow p_{ji} = \|\rho_j\, L(e_i)\|,$$

$$\rho_j\, L^2(e_i) = \rho_j\Big(\sum_{k,t} p_{tk} p_{ki}\, e_t\Big) = \Big(\sum_k p_{jk} p_{ki}\Big) e_j,$$

then we have

$$\big\|\rho_j\, L^2(e_i)\big\| = \sum_k \|\rho_j\, L(e_k)\| \cdot \|\rho_k\, L(e_i)\|.$$

Therefore, we have a 2-step Chapman–Kolmogorov equation in probability theory,

$$p_{ji}^{(2)} = \big\|\rho_j\, L^2(e_i)\big\| = \sum_k p_{jk} p_{ki}.$$

For the $(l+m)$-step, we use the matrix representation of L^{l+m} that is A^{l+m}. We have

$$p_{ji}^{(l+m)} = \big\|\rho_j\, L^{l+m}(e_i)\big\| = \begin{pmatrix} 0 & \cdots & 0\ 1\ 0 & \cdots & 0 \end{pmatrix} A^{l+m} \begin{pmatrix} 0 \\ \vdots \\ 0 \\ 1 \\ 0 \\ \vdots \\ 0 \end{pmatrix} i$$

$$= \sum_{t_1 \ldots t_{l+m-1}} a_{jt_1} a_{t_1 t_2} \cdots\cdots a_{t_{l+m-1} i}$$

$$= \sum_{t_1 \ldots t_{l+m-1}} a_{j\, t_1} \cdots a_{t_{l-1} k}\, a_{kt_{l+1}} \cdots\cdots a_{t_{l+m-1} i}$$

$$= \sum_k p_{jk}^{(l)} \cdot p_{ki}^{(m)}$$

$$= \sum_k \|\rho_j\, L^l(e_k)\| \cdot \|\rho_k\, L^m(e_i)\|.$$

Thus, we verified our version of the Chapman–Kolmogorov equation. As to the version $L^{(r+n+m,\ m)} = L^r \circ L^{(n+m,\ m)}$, it is easy to see, since we run the chain again when it has already moved m steps. Thus, the Chapman–Kolmogorov equation in evolution algebras is an operator equation.

Remark 3. As we see, in the evolution algebra corresponding to a given Markov chain, probabilities, as an interpretation of coefficients of elements, can be found by using the evolution operator and projections. For example,

$$\rho_j\, L(e_i) = p_{ji} e_j,$$

$$\rho_j\, L^n(e_i) = p_{ji}^{(n)} e_j.$$

They can be used to find some useful relations between Markov chains and their corresponding evolution algebras.

4.1.4 Concepts related to evolution operators

We need some concepts about different types of elements in an evolution algebra and different types of evolution operators, such as nonnegative elements, negative elements, nonpositive elements and positive elements, positive evolution operators, nonnegative evolution operators and periodical positive evolution operators, etc. Let us now define them here.

Definition 7. *Let $x = \sum_i a_i e_i$ be an element in the evolution algebra M_X that corresponds to a Markov chain X. We say x is a nonnegative element if a_i, $i \in \Lambda$, are all nonnegative elements in field K. If a_i are all negative, we say x is negative. If a_i are all positive, we say x is positive. If a_i are all nonpositive, we say x is nonpositive.*

Definition 8. *For any nonnegative element $x \neq 0$, if $L(x)$ is positive, we say L is positive; if $L(x)$ is nonnegative, we say L is nonnegative. If L is nonnegative, and for any generator e_i, $\rho_i L(e_i) \neq 0$ periodically occurs, we say L is periodically positive.*

Lemma 5. *For a nonnegative or nonpositive element x, we have $\|L(x)\| \leq \|x\|$.*

Proof. Let $x = \sum_i a_i e_i$, then $L(x) = \sum_i a_i L e_i = \sum_i a_i p_{ki} e_k$. $\|L(x)\| = |\sum_i a_i p_{ki}| \leq |\sum_i a_i \sum_k p_{ki}| \leq |\sum_i a_i| = \|x\|$.

4.1.5 Basic algebraic properties of Markov chains

Markov chains have many interesting algebraic properties as we will see in this chapter. Here let us first present several basic propositions.

Theorem 17. *Let C be a subset of the state set $S = \{e_i \mid i \in \Lambda\}$ of a Markov chain X. C is closed in the sense of probability if and only if C generates an evolution subalgebra of the evolution algebra M_X.*

Proof. By the definition of closed subset of the state set in probability theory, C is closed if and only if for all states e_i and e_j, $e_j \in C$, $e_i \notin C$, $p_{i\,j} = 0$, which just means

$$e_j \cdot e_j = \sum_i p_{ij} e_i = \sum_{e_k \in C} p_{kj} e_k.$$

Then, if we denote the subalgebra that is generated by C by $\langle C \rangle$, it is clear that $e_j \cdot e_j \in \langle C \rangle$, whenever $e_j \in C$. Thus, C generates an evolution algebra.

Corollary 11. *If a subset C of the state set $S = \{e_i \mid i \in \Lambda\}$ of the Markov chain X is closed, then $\rho_j L^n(e_i) = 0$ for $e_i \in C$ and $e_j \notin C$.*

Proof. Since C generates an evolution subalgebra and the evolution operator leaves a subalgebra invariant, $L^n(e_i) \in C$ for any $e_i \in C$ and any positive integer n. That is, any projection to the out of the subalgebra $\langle C \rangle$ is zero. Particularly, $\rho_j L^n(e_i) = 0$. In term of probability, $p_{ji}^{(n)} = 0$.

In Markov chains, a closed subset of the state set is referred as the impossibility of escaping. That is, a subset C is closed if the chain once enters C, it can never leave C. In evolution algebras, a subalgebra has a kind of similar significance. A subalgebra generated by a subset C of the generator set is closed under the multiplication. That is, there is no new generator that is not in C that can be produced by the multiplication. Furthermore, the evolution operator leaves a subalgebra invariant.

Corollary 12. *State e_k is an absorbing state in the Markov chain X if and only if e_k is an idempotent element in the evolution algebra M_X.*

Proof. State e_k is an absorbing state in Markov chain X if and only if $p_{kk} = 1$. So, in the algebra M_X, we have $e_k \cdot e_k = e_k$.

Remark 4. If e_k is an absorbing state, then for any positive integer n, $L^n(e_k) = e_k$ and e_k generates a subalgebra with dimension one, $\langle e_k \rangle = R e_k$, where R is the real number field.

Theorem 18. *A Markov chain X is irreducible if and only if the corresponding evolution algebra M_X is simple.*

Proof. If M_X has a proper evolution subalgebra A with the generator set $\{e_i \mid i \in \Lambda_0\}$, then extend this set to a natural basis for M_X as $\{e_i \mid i \in \Lambda\}$, where $\Lambda_0 \subseteq \Lambda$. For any $i \in \Lambda_0$, since $e_i \cdot e_i = \sum_{k \in \Lambda_0} p_{ki} e_k$, so for any $j \notin \Lambda_0$, $p_{ji} = 0$. That is, $\{e_i \mid i \in \Lambda_0\}$ is closed in the sense of probability, which means the Markov chain M is not irreducible.

On the other hand, if the Markov chain X is not irreducible, the state set $S = \{e_i \mid i \in \Lambda\}$ has a proper closed subset in the sense of probability. As Theorem 17 shows, M_X has a proper evolution subalgebra.

4.2 Algebraic Persistency and Probabilistic Persistency

In this section, we discuss the difference between algebraic concepts, algebraic persistency and algebraic transiency, and analytic concepts, probabilistic persistency and probabilistic transiency. When the dimension of the evolution algebra determined by a Markov chain is finite, algebraic concepts and analytic concepts are equivalent. By "equivalent" we means that, for example, a generator is algebraically persistent if and only if it is probabilistically persistent. Generally, a generator is probabilistically transient if it is algebraically transient, and a generator is algebraically persistent if it is probabilistically persistent. To this end, we need to define destination operators and other algebraic counterparts of concepts in probability theory.

4.2.1 Destination operator of evolution algebra M_X

Definition 9. *Denote $\rho_j^o = \sum_{k \neq j} \rho_k$. We call ρ_j^o the deleting operator, which deletes the component of e_j, i.e., $\rho_j^o(x) = x - \rho_j(x)$. Then, we can define operators of the first visiting to a generator (characteristic state) e_j as follows:*

$$V^{(1)} = \rho_j L, \text{ it happens at the first time,}$$

$$V^{(2)} = V^{(1)} \rho_j^o L, \text{ it happens at the second time,}$$

$$V^{(3)} = V^{(2)} \rho_j^o L, \text{ happens at the third time,}$$

$$\dots\dots\dots\dots\dots\dots\dots\dots\dots\dots\dots$$

$$V^{(m)} = V^{(m-1)} \rho_j^o L, \text{ it happens at the m-th time,}$$

we define a destination operator (notice, e_j is a "destination"):

$$D_j = \sum_{m=1}^{\infty} V^{(m)}$$

$$= \sum_{m=1}^{\infty} \rho_j L \left(\rho_j^o L \right)^{(m-1)}.$$

Lemma 6. *The destination operator D_i is convergent.*

Proof. Since $D_i = \sum_{m=1}^{\infty} \rho_i L \left(\rho_i^o L \right)^{(m-1)} = \rho_i L \sum_{m=1}^{\infty} \left(\rho_i^o L \right)^{(m-1)}$, when consider operator $\rho_i^o L$ under the natural basis, we have a matrix representation for $\rho_i^o L$, denote this matrix by A. Then, A is the matrix obtained from the matrix representation of L by replacing its ith row by zero row. Explicitly,

$$A = \begin{bmatrix} p_{11} & p_{12} & p_{13} & \cdots \\ \cdots & \cdots & \cdots & \cdots \\ p_{i-1,1} & p_{i-1,2} & p_{i-1,3} & \cdots \\ 0 & 0 & 0 & \cdots \\ p_{i+1,1} & p_{i+1,2} & p_{i+1,3} & \cdots \\ \vdots & \vdots & \vdots & \vdots \end{bmatrix}.$$

If we define a norm for matrices $B = (b_{ij})$ to be $\|B\| = \max_j\{\sum_i |b_{ij}|\}$, then, it is easy to check that the norm of operator $\rho_i^o L$ is the maximum of the summation of absolute values of entries in each column of A. That is,

$$\|\rho_i^o L\| = \|A\| = \max\{\sum_k p_{kj} \mid j \in \Lambda\}.$$

Case I.   If all $p_{ik} = 0$, $k \in \Lambda$, then

$$\rho_i L(e_k) = 0, \quad \rho_i L(\rho_i^o L)(e_k) = 0, \cdots,$$

then

$$D_i(e_k) = 0, \quad \forall\, k \in \Lambda.$$

Case II.   Not all $p_{i1}, p_{i2}, \cdots, p_{in}, \cdots$ are zero, then $\|A^{k_0}\| \le r_0 < 1$ for some integer k_0, since no column in A^{k_0} sums to 1. Then $\|A^n\| \le r_0^{\lceil \frac{n}{k_0} \rceil} < 1$. Since A or $\rho_i^o L$ belongs to the normed algebra $L(M)$, we can utilize theorems in Functional Analysis. Thus, we get the existence of the limit $\lim_{n \to \infty} \sqrt[n]{\|A^n\|}$. Then, we set $\lim_{n \to \infty} \sqrt[n]{\|A^n\|} = r < 1$ or $\lim_{n \to \infty} \sqrt[n]{\|(\rho_i^o L)^n\|} = r$.

Claim:

$$(I - \rho_i^o L)^{-1} = \sum_{n=0}^{\infty} (\rho_i^o L)^n.$$

Since for any $\epsilon > 0$ and $r + \epsilon < 1$, there is $N > k_0$, for $n \ge N$

$$\sqrt[n]{\|A^n\|} = \sqrt[n]{\|(\rho_i^o L)^n\|} < r + \epsilon,$$

so

$$\|(\rho_i^o L)^n\| < (r + \epsilon)^n.$$

We have, for $m > N$

$$\left\| \sum_{n=m}^{\infty} (\rho_i^o L)^n \right\| \le \sum_{n=m}^{\infty} \|A^n\| \le \sum_{n=m}^{\infty} (r + \epsilon)^n = \frac{(r + \epsilon)^m}{1 - r - \epsilon}.$$

Therefore, $\sum_{n=0}^{\infty} (\rho_i^o L)^n$ converges by norm. Denote $B = \sum_{n=0}^{\infty} (\rho_i^o L)^n$, we need to check

$$B(I - \rho_i^o L) = (I - \rho_i^o L)B = I.$$

Set

$$B_m = \sum_{n=0}^{m} (\rho_i^o L)^n$$

then

$$B_m(I - \rho_i^o L) = B_m - B_m(\rho_i^o L)$$
$$= (I - \rho_i^o L)B_m = I - (\rho_i^o L)^{m+1}.$$

But $||B_m - B|| \longrightarrow 0$, when $m \geq N$, we have

$$\left\|(\rho_i^o L)^{m+1}\right\| \leq (r + \epsilon)^{m+1} \longrightarrow 0,$$

then we get

$$B(I - \rho_i^o L) = (I - \rho_i^o L)B = I.$$

Thus

$$D_i = \rho_i L \sum_{m=1}^{\infty} (\rho_i^o L)^{m-1} = \frac{\rho_i L}{I - \rho_i^o L},$$

which means that the operator D_i converges.

Corollary 13. $\|D_i(e_k)\| \leq 1$.

Proof. From the proof of the above Lemma 6, we see that
in case I,

$$\|D_i(e_k)\| = 0;$$

in case II,

$$\|I - \rho_i^o L\| \geq 1,$$

since $\|I - A\| \geq 1$ (because of (i, i)−entry of $(I - A)$ is 1) and $\|\rho_i L\| \leq 1$.
Then $\|D_i(e_k)\| \leq 1$.

Lemma 7. $\rho_j L^n = \sum_{k=1}^{n} \rho_j L^{n-k} \left(\rho_j L \left(\rho_j^o L\right)^{k-1}\right).$

Proof. We use induction to prove this lemma. When $n = 1$, $\rho_j L = \rho_j(\rho_j L)$.
Suppose when $n = n$, the formula is correct. Then, since

$$L = \left(\rho_j + \rho_j^o\right)L = \rho_j L + \rho_j^o L,$$

we have

$$\rho_j L^{n+1} = \rho_j L^n L$$

$$= \sum_{k=1}^{n} \rho_j L^{n-k} \left(\rho_j L \left(\rho_j^o L\right)^{k-1}\right)\left(\rho_j L + \rho_j^o L\right)$$

$$= \sum_{k=1}^{n} \rho_j L^{n-k} \left(\rho_j L \left(\rho_j^o L\right)^{k-1}\right)\left(\rho_j L\right) + \sum_{k=1}^{n} \rho_j L^{n-k} \left(\rho_j L \left(\rho_j^o L\right)^{k}\right)$$

$$= \rho_j L^n \left(\rho_j L\right) + \sum_{k=1}^{n} \rho_j L^{n-k} \left(\rho_j L \left(\rho_j^o L\right)^{k}\right)$$

$$= \sum_{k=1}^{n+1} \rho_j L^{n+1-k} \left(\rho_j L \left(\rho_j^o L\right)^{k-1}\right).$$

Thus, we got the proof.

Theorem 19. $\|Q_j(e_j)\| = \frac{1}{1-\|D_j(e_j)\|}$, where $Q_j = \sum_{n=0}^{\infty} \rho_j L^n$.

Proof. By utilizing the Lemma 7, we have

$$Q_j(e_j) = \rho_j(e_j) + \sum_{n=1}^{\infty} \rho_j L^n(e_j)$$

$$= e_j + \sum_{n=1}^{\infty} \left(\sum_{k=1}^{n} \rho_j L^{n-k} \left(\rho_j L \left(\rho_j^o L \right)^{k-1} \right) \right)$$

$$= e_j + \sum_{n=1}^{\infty} \sum_{k=1}^{n} \left\| \rho_j L \left(\rho_j^o L \right)^{k-1}(e_j) \right\| \rho_j L^{n-k}(e_j)$$

$$= e_j + \sum_{k=1}^{\infty} \sum_{n=k}^{\infty} \left\| \rho_j L \left(\rho_j^o L \right)^{k-1}(e_j) \right\| \rho_j L^{n-k}(e_j).$$

In the last step, we have utilized Fubini's theorem. Thus, we have

$$\|Q_j(e_j)\| = 1 + \sum_{k=1}^{\infty} \sum_{n=k}^{\infty} \left\| \rho_j L \left(\rho_j^o L \right)^{k-1}(e_j) \right\| \left\| \rho_j L^{n-k}(e_j) \right\|$$

$$= 1 + \|D_j(e_j)\| \|Q_j(e_j)\|.$$

Therefore, we get

$$\|Q_j(e_j)\| = \frac{1}{1 - \|D_j(e_j)\|}.$$

Theorem 20. *If $D_j(e_j) = e_j$, then the generator e_j as a characteristic state is persistent in the sense of probability.*

If $D_j(e_j) = ke_j$, $0 \le k < 1$, then the generator e_j as a characteristic state is transient in the sense of probability.

Proof. By comparing our definition of the first visiting operators with the first visits to some state in Markov chain theory, we can find that the coefficient of $\rho_j L \left(\rho_j^o L \right)^{m-1}(e_i)$ is the probability that the first visit to state e_j from e_i, which is $f_{ij}^{(m)}$ in Probability theory. Therefore, our statement is correct in the sense of probability.

Corollary 14. *In the sense of probability, generator e_j as a characteristic state is persistent if and only if $\|Q_j(e_j)\| = \infty$, and e_j is transient if and only if $\|Q_j(e_j)\| < \infty$.*

Proof. By Theorem 20, e_j is persistent in probability if and only if $\|D_j(e_j)\|=1$, then using Theorem 19, we get e_j is persistent if and only if $\|Q_j(e_j)\| = \infty$. Similarly, we can get the second statement in the corollary.

We now say e_j is **probabilistically persistent** if it is persistent in the sense of probability, and e_j is **probabilistically transient** if it is transient in the sense of probability.

4.2.2 On the loss of coefficients (probabilities)

Lemma 8. *If $\rho_j L^{n_0}(e_i) \neq 0$, $i \neq j$, and n_0 is the least number that has this property, then $\rho_j (\rho_i^0 L)^{n_0}(e_i) \neq 0$.*

Proof. If $n_0 = 1$, this is obvious.

If $n_0 > 1$, since L is a linear map, $\rho_j L(e_j) = 0$, but, $\rho_j L^{n_0}(e_i) \neq 0$, then e_j must come from some element e_k which is not e_i. So each time when the action of L is taken, we delete e_i, which does not affect the final result.

Proposition 8. *If there is e_j that occurs in $\langle e_i \rangle$, such that e_i does not occur in $\langle e_j \rangle$, then $D_i(e_i) = k e_i$, $k < 1$. That is, e_i is transient in the sense of probability. There is a loss of probability, $1 - k$.*

Proof. Since e_j occurs in $\langle e_i \rangle$, so $\rho_j L^{n_0}(e_i) \neq 0$, for some n_0. e_i does not occur in $\langle e_j \rangle$, so $\rho_i L^k(e_j) = 0$, for any integer k.

If $n_0 = 1$, $\rho_j L(e_i) = p_{ji} e_j \neq 0$. We see

$$D_i = \sum_{m=1}^{\infty} \rho_i L(\rho_i^0 L)^{m-1} = \rho_i \sum_{m=1}^{\infty} (L\rho_i^0)^{m-1} L = \rho_i T_i L,$$

where

$$T_i = \sum_{m=1}^{\infty} (\rho_i^0 L)^{m-1}.$$

Then, we compute

$$
\begin{aligned}
D_i(e_i) &= \rho_i T_i L(e_i) \\
&= \rho_i T_i \left(p_{ii} e_i + p_{ji} e_j + \sum_{k \neq i, k \neq j} p_{ki} e_k \right) \\
&= p_{ii} e_i + p_{ji} \rho_i T_i (e_j) + \sum_{k \neq i, k \neq j} p_{ki} \rho_i T_i (e_k).
\end{aligned}
$$

As the proof of the convergence of the destination operator in Lemma 6, we have

$$T_i = (I - L\rho_i^0)^{-1},$$

and

$$\|\rho_i T_i (e_k)\| \leq 1.$$

Since $\rho_i L^k(e_j) = 0$, so then $\rho_i T_i(e_j) = 0$. Therefore

$$\|D_i(e_i)\| \leq p_{ii} + \sum_{k \neq i, k \neq j} p_{ki} \leq 1 - p_{ji}.$$

If $n_0 > 1$, we derive

$$D_i = \sum_{m=1}^{n_0-1} \rho_i \, L \, (\rho_i^o \, L)^{m-1} + \rho_i \, L \, (\rho_i^o \, L)^{n_0-1} + \rho_i \, L \, (\rho_i^o \, L)^{n_0} + \cdots\cdots$$

$$= \sum_{m=1}^{n_0-1} \rho_i \, L \, (\rho_i^o \, L)^{m-1} + \rho_i \, L \left(\sum_{k=1}^{\infty} (\rho_i^o \, L)^{k-1} \right) (\rho_i^o \, L)^{n_0-1}$$

$$= \sum_{m=1}^{n_0-1} \rho_i \, L \, (\rho_i^o \, L)^{m-1} + \rho_i \, T_i \, L \, (\rho_i^o \, L)^{n_0-1}$$

$$= A + \rho_i \, T_i \, L \, (\rho_i^o \, L)^{n_0-1},$$

where $A = \sum_{m=1}^{n_0-1} \rho_i \, L \, (\rho_i^o L)^{m-1}$. Then, acting on e_i, we have

$$D_i(e_i) = A(e_i) + \rho_i T_i L(\rho_i^o L)^{n_0-1}(e_i)$$

$$= A(e_i) + \rho_i T_i \left(a e_j + \sum_{k \in \Lambda_1} a_k e_k \right)$$

$$= A(e_i) + a\rho_i T_i(e_j) + \sum_{k \in \Lambda_1} a_k \rho_i T_i(e_k),$$

where, $a > 0$, Λ_1 is a proper index subset. Since $\|A(e_i)\| + a + \sum_{k \in \Lambda_1} |a_k| \leq 1$, so $\|A(e_i)\| \neq 1$. But, $\rho_i T_i (e_j) = 0$, therefore $\|D_i(e_i)\| \leq 1 - a$. Thus, e_i is transient in the sense of probability. There is a loss of probability, $1 - k$. Thus, we finish the proof.

Lemma 9. *Generator e_i is transient in the algebra M_X if and only if there is e_j which occurs in $\langle e_i \rangle$, such that e_i does not occur in $\langle e_j \rangle$.*

Proof. Because e_j occurs in $\langle e_i \rangle$, by the definition of an evolution subalgebra, $e_j \in \langle e_i \rangle$. So, $\langle e_j \rangle \subset \langle e_i \rangle$. But, e_i does not occur in $\langle e_j \rangle$. This means $\langle e_i \rangle$ does not contain in $\langle e_j \rangle$. Therefore, $\langle e_i \rangle$ has a proper subalgebra. By definition, e_i is transient in the algebra M_X. On the other hand, if e_i is transient in M_X, $\langle e_i \rangle$ is not a simple algebra. It must have a proper evolution subalgebra, for example, $E \subset \langle e_i \rangle$. Then, E has a natural basis that can be extended to a natural basis of $\langle e_i \rangle$. Since e_i belongs to the natural basis of $\langle e_i \rangle$, so there must be an e_j in the basis of E. Thus, e_i does not occur in $\langle e_j \rangle$.

From Proposition 8 and Lemma 9, if a generator e_i is algebraically transient, then it is also probabilistically transient.

Theorem 21. *Let M be a finite dimensional evolution algebra. If $D_i(e_i) = ke_i$, $0 \leq k < 1$, then there exists e_j which occurs in $\langle e_i \rangle$, but e_i does not occur in $\langle e_j \rangle$.*

Proof. Suppose that for all e_j that occurs in $\langle e_i \rangle$, e_i also occurs in $\langle e_j \rangle$. Then for convenience, we assume $e_1, e_2, \cdots, e_i, \cdots e_t$ are all generators which occur in $\langle e_i \rangle$, and $e_i < \langle e_j \rangle$, $j = 1, 2, \cdots, t$. We consider evolution subalgebras $\langle e_i \rangle$

and all $\langle e_j \rangle$, we must have $\langle e_i \rangle = \langle e_j \rangle$, $j = 1, 2, \cdots, t$. This means $\langle e_i \rangle$ is an irreducible evolution subalgebra.

Case 1. If e_i is aperiodic, for simplicity, we take

$$L(e_i) = a_1 e_1 + a_2 e_2 + \cdots + a_t e_t,$$

where $0 < a_j < 1$ and $\sum_{j=1}^{t} a_j = 1$. That is, $\rho_i L(e_j) = p_{ij} e_i \neq 0$ for any pair (i, j). Otherwise, we start from some power of L. Now, let us look at

$$\rho_i L^2(e_i) = (a_1 p_{i1} + a_2 p_{i2} + \cdots + a_t p_{it})e_i,$$

and denote

$$c = \min\{p_{i1}, p_{i2}, \cdots, p_{it}\}.$$

Since $a_1 p_{i1} + a_2 p_{i2} + \cdots + a_t p_{it}$ is the mean of $p_{i1}, p_{i2}, \cdots, p_{it}$ (because of $\sum_{j=1}^{t} a_j = 1$), so $\sum_{j=1}^{t} a_k p_{ik} \geq c$. That is, $\|\rho_i L^2(e_i)\| \geq c$. Set $L^2(e_i) = A_1 e_1 + A_2 e_2 + \cdots + A_t e_t$. Since L^2 preserves the norm, so $A_1 + A_2 + \cdots + A_t = 1$, and $0 < A_j < 1$. Look at

$$\rho_i L^3(e_i) = (A_1 p_{i1} + A_2 p_{i2} + \cdots + A_t p_{it})e_i,$$

Then, $\|\rho_i L^3(e_i)\| = \sum_{k=1}^{t} A_k p_{ik} \geq c$. Inductively, we have $\|\rho_i L^n(e_i)\| \geq c$, $(n > 1)$. This just means that $\|\rho_i L^n(e_i)\|$ does not approach to zero, thus

$$\sum_{n=1}^{\infty} \|\rho_i L^n(e_i)\| = \infty.$$

Therefore, we have $D_i(e_i) = e_i$, which contradicts $D_i(e_i) = k e_i$, where $0 \leq k < 1$.

Case 2. If $\langle e_i \rangle$ is periodical with a period of d. We consider operator L^d. Since L^d can be written as a direct sum $L^d = l_0 \oplus l_1 \oplus \cdots \oplus l_{d-1}$. Consequently $\{e_1, e_2, \cdots, e_t\}$ has a partition with d cells. Suppose e_i is in subspace Δ_k, which is spanned by the kth cell of the partition, then we consider l_k. Similarly, we will have $\|\rho_i l_k^n(e_i)\| > 0$. Because $\sum_{n=1}^{\infty} \|\rho_i l_k^n(e_i)\|$ is a sub-series of $\sum_{n=1}^{\infty} \|\rho_i L^n(e_i)\|$, so we still get $\sum_{n=1}^{\infty} \|\rho_i L^n(e_i)\| = \infty$. $(\sum_{n=1}^{\infty} \|\rho_i L^n(e_i)\| \geq \sum_{n=1}^{\infty} \|\rho_i l_k^n(e_i)\| = \infty)$. We finish the proof.

Theorem 22. *(A generalized version of theorem 21) Let $D_i(e_i) = k e_i$, $0 \leq k < 1$. When $\langle e_i \rangle$ is a finite dimensional evolution subalgebra, then there exists e_j which occurs in $\langle e_i \rangle$, but e_i does not occur in $\langle e_j \rangle$.*

Remark 5. Let's summarize that when $\langle e_i \rangle$ is a finite dimensional evolution subalgebra, e_i is algebraically transient if and only if e_i is probabilistically transient. Now we can use this statement to classify states of a Markov chain. In Markov Chain theory, it is not easy to check if a state e_i is transient, while in evolution algebra theory, it is easy to check if e_i is algebraically transient.

4.2.3 On the conservation of coefficients (probabilities)

We work on Markov evolution algebras, for example, M_X, which has a generator set $\{e_i : i \in \Lambda\}$.

Lemma 10. *Generator e_i is algebraically persistent if and only if all generators e_j which occurs in $\langle e_i \rangle$, e_i also occurs in $\langle e_j \rangle$.*

Proof. If e_j occurs in $\langle e_i \rangle$, then subalgebra $\langle e_j \rangle \subseteq \langle e_i \rangle$. Since $\langle e_i \rangle$ is a simple evolution subalgebra, so we have $\langle e_j \rangle = \langle e_i \rangle$. That is, e_i must occur in $\langle e_j \rangle$. On the other hand, if $\langle e_i \rangle$ is not a simple evolution subalgebra, it must have a proper subalgebra, say B. Then, B has a natural basis that can be extended to the natural basis of $\langle e_i \rangle$. Let e_k be a generator in B, then e_i does not occur in $\langle e_k \rangle$.

Lemma 11. *Let M_X is a finite dimensional evolution algebra. If for all generators e_j which occurs in $\langle e_i \rangle$, e_i also occurs in $\langle e_j \rangle$, then $D_i(e_i) = e_i$. That is, if e_i is algebraically persistent, then e_i is also probabilistically persistent.*

Proof. If e_i is not probabilistically persistent, that is $D_i(e_i) = ke_i$, where $0 \leq k < 1$, then by Theorem 22, there exists some e_j that occurs in $\langle e_i \rangle$. But e_i does not occur in $\langle e_j \rangle$. Thus $\langle e_j \rangle \subseteq \langle e_i \rangle$, so $\langle e_i \rangle$ is not simple.

Theorem 23. *If e_i is probabilistically persistent, then e_i is algebraically persistent, i.e., for any e_j which occurs in $\langle e_i \rangle$, e_i also occurs in $\langle e_j \rangle$.*

Proof. If e_i is not algebraically persistent, e_i is algebraically transient. By Proposition 8, we have $D_i(e_i) = ke_i$ with $0 \leq k < 1$.

Remark 6. Let us summarize that when $\langle e_i \rangle$ is a finite dimensional evolution subalgebra, e_i is algebraically persistent if and only if e_i is probabilistically persistent. In Markov Chain theory, we have to compute a series of probabilities in order to check if a state e_i is persistent; while in evolution algebra theory, it is easy to check if the subalgebra $\langle e_i \rangle$ generated by e_i is simple. As the remark in the last subsection, we can use this statement to classify states of a Markov chain.

Theorem 24. *An evolution algebra is simple if and only if each generator that occurs in the evolution subalgebra can be generated by any other generator.*

Proof. If e_{i_0} does not occur in certain $\langle e_{j_0} \rangle$, then $\langle e_{j_0} \rangle$ is a proper subalgebra of the evolution algebra. But it is irreducible, which is a contradiction. If the evolution algebra is not simple, then it has a proper subalgebra, say A. There is a generator of the algebra, for example e_{i_0}, e_{i_0} does not occur in A. So there is another generator e_j of the algebra A, such that e_{i_0} does not occur in $\langle e_j \rangle$. This is a contradiction.

Theorem 25. *For any finite state Markov chain, there is always a persistent state.*

Proof. This is a consequence of Theorem 9 in Chapter 3.

Proposition 9. *All generators in the same simple evolution algebra (or subalgebra) M_X are of the same type with respect to periodicity and persistency. That is, in the same closed subset of the state space, all states are of the same type with respect to periodicity and persistency.*

Proof. This is a consequence of Theorem 7, 8, and Corollary 9 in Chapter 3.

Remark 7. The above Theorem 24 characterizes a simple evolution algebra, namely, characterizes an irreducible Markov chain. However, we do not have this kind of simple characteristics in Markov chain theory as a counterpart. It provides an easy way to verify irreducible Markov chains.

We see from Chapter 3, the proof of Theorem 9 is quite easy. However, it is a laborious work to prove Theorem 25 in Markov chain theory.

The same remark for the proof of Proposition 9 as that for Theorem 25 is true. They all show that evolution algebra theory has some advantages in study classical theory as the study of Markov chains.

4.2.4 Certain interpretations

- If an evolution algebra M_X is connected, then in its corresponding Markov chain, for any pair of the states, there is at least one sequence of states that can be accessible from the other (but may not be necessarily two-way accessibility).
- A semisimple evolution algebra is not connected. For an evolution algebra M_X, the probabilistic meaning of this statement is that a semisimple evolution algebra corresponds to a collection of several Markov chains that are independent. The number of these independent Markov chains is the number of components of the direct sum of the semisimple evolution algebra.
- Interpretation of Theorem 8 in Chapter 3: Let e_i and e_j be elements in a natural basis of an evolution algebra. If e_i and e_j can intercommunicate and both are algebraically persistent, then they belong to the same simple evolution subalgebra of M_X, which means, e_i and e_j belong to the same closed subset of the state space.
- Interpretation of Corollary 9 in Chapter 3, for finite dimensional evolution algebra, we have the following statements.

1). A finite state Markov chain X has a proper closed subset of the state space if and only if it has at least one transient state.

2). A Markov chain X is irreducible if and only if it has no transient state.

3). If a Markov chain X has no transient state, then it is irreducible or it is a collection of several independent irreducible Markov chains.

4.2.5 Algebraic periodicity and probabilistic periodicity

In the section 3.4.1 of Chapter 3, plenary powers are used to define (algebraically) periodicity. An equivalent definition of periodicity was given by using evolution operators. When considering the matrix representation of an evolution operator, we can see that the algebraic definition is the same as the probabilistic one. Therefore, we have the following statement.

Proposition 10. *For a generator in an evolution algebra M_X, its algebraic periodicity is the same as its probabilistic periodicity.*

4.3 Spectrum Theory of Evolution Algebras

In this section, we study the spectrum theory of the evolution algebra M_X determined by a Markov chain X. Although the dynamical behavior of an evolution algebra is embodied by various powers of its elements, the evolution operator seems to represent a "total" principal power. From the algebraic viewpoint, we study the spectrum of an evolution operator. Particularly, an evolution operator is studied at the 0th level in its hierarchy of the evolution algebra, although we do not study it at high level, which would be an interesting further research topic. Another possible spectrum theory could be a study of the plenary powers. Actually, we have already defined plenary powers for a matrix in the proof of Proposition 7 in Chapter 3. It could be a way to study this possible spectrum theory.

4.3.1 Invariance of a probability flow

We give a proposition to state our point first.

Proposition 11. *Let L be the evolution operator of the evolution algebra M_X corresponding to the Markov chain X, then for any nonnegative element y, $\|L(y)\| = \|y\|$.*

Proof. Write $y = \sum_{i=1}^{n} a_i e_i$, then $L(y) = \sum_{i=1}^{n} \sum_{k=1}^{n} p_{ik} a_k e_i$. Therefore

$$\|L(y)\| = \left\| \sum_{i=1}^{n} \sum_{k=1}^{n} p_{ik} a_k e_i \right\|$$
$$= \sum_{i=1}^{n} \sum_{k=1}^{n} p_{ik} a_k$$
$$= \sum_{k=1}^{n} a_k = \|y\|.$$

As we see, a Markov chain, as being a dynamical system, preserves the total probability flow. Suppose we start at a general state y with the total probability $\|y\|$. After one step motion, the total probability is still $\|y\|$. Because of this kind of conservation or invariance of flow, it is easy to understand the so-called equilibrium states as the following theorem states.

Theorem 26. *For any nonnegative, nonzero element x_0 in the evolution algebra M_X determined by Markov chain X, there is an element y in M_X so that $L(y) = y$ and $\|y\| = \|x_0\|$, where L is the evolution operator of M_X.*

Proof. We assume the algebra is finite dimensional. Set

$$D_{x_0} = \left\{ \sum_{i=1}^{n} a_i e_i \mid 0 \leq a_i \leq \|x_0\|, \sum_{i=1}^{n} a_i = \|x_0\| \right\}.$$

Then D_{x_0} is a compact subset and $L(D_{x_0}) \subseteq D_{x_0}$. Since L is continuous, we can use Brouwer's fixed point theorem to get a fixed point y. All we need to observe is that the fixed point is also in D_{x_0}, so then $\|y\| = \|x_0\|$.

Symmetrically, we may consider a nonpositive, nonzero element x_0 to get a fixed point. If consider the unit sphere D in the Banach space M_X, we can get an equilibrium state by this theorem. On the other hand, L, as a linear map, has eigenvalue 1 as the theorem showed. We state a theorem here.

Theorem 27. *Let M_X be an evolution algebra with dimension n, then the evolution operator L has eigenvalue 1 and 1 is an eigenvalue that has the greatest absolute value.*

Proof. By Theorem 26, L has a fixed point y, $y \neq 0$. Since L is linear, $L(0) = 0$. So we take y as a vector. Then $L(y) = y$ means 1 is an eigenvalue of L. If λ is any other eigenvalue, x is an eigenvector that corresponds to λ, then $L(x) = \lambda x$. We know $\|L(x)\| \leq x$, which is $\|\lambda x\| \leq \|x\|$. Thus, we obtain $\|\lambda\| \leq 1$.

4.3.2 Spectrum of a simple evolution algebra

Simple evolution algebras can be categorized as periodical simple evolution algebras and aperiodic simple evolution algebras. Consequently, their evolution operators can also be grouped as positive evolution operators and periodical evolution operators. The notion, positive evolution operator here, is slightly general. Let us first give the definition.

Definition 10. *Let L be the evolution operator of the evolution algebra M_X corresponding to the Markov chain X. We say L is positive if there is a positive integer m for any generators e_i and e_j, we have*

$$\rho_j L^m(e_i) \neq 0.$$

Theorem 28. *Let L be a positive evolution operator of an evolution algebra, then the geometric multiplicity corresponding to the eigenvalue one is 1.*

Proof. Since L is positive, there is an integer m such that for any pair e_k, e_l, we have $\rho_k L^m(e_l) \neq 0$. Consider L is a continuous map from D to itself. Assume L has two fixed points x_0, y_0 and $x_0 \neq \lambda y_0$. Since L is linear, $L(0) = 0$, so we can take x_0, y_0 as vectors $\overrightarrow{X_0}$, $\overrightarrow{Y_0}$ from the original 0 to x_0 and y_0, respectively. Then the subspace M_1 spanned by $\overrightarrow{X_0}$ and $\overrightarrow{Y_0}$ will be fixed by L.

Case I. If this evolution algebra is dimension 2, then L fixes the whole underlying space of the algebra. That means $L(e_1) = e_1$ and $L(e_2) = e_1$. Therefore $\rho_2 L(e_1) = 0$ and $\rho_1 L(e_2) = 0$. This is a contradiction.

Case II. If the dimension of M_X is greater than 2, then $M_1 \cap (\partial D_0) \neq \phi$, where $D_0 = \{\sum_{i=1}^n a_i e_i \mid 0 \leq a_i \leq 1, \sum_{i=1}^n a_i \leq 1\}$. Since x_0, $y_0 \in D_0$, and L is linear, so the line l that passes through x_0 and y_0 will be fixed by L. $l \subset M_1$ and $l \cap D \neq \phi$, for any $z \in l \cap D$. Writing z as $z = \sum_{i=1}^n a_i e_i$, there must be some a_i that is equal to 0, say $a_n = 0$. Then, because $L^m(z) = z$, $(L(z) = z)$, we have $\rho_n L^m(z) = \rho_n(z) = 0$. This is a contradiction.

Thus, the eigenspace of the eigenvalue one has to be dimension 1.

Theorem 29. *If M_X is a finite dimensional simple aperiodic evolution algebra, its evolution operator is positive.*

Proof. Let the generator set of M_X be $\{e_1, e_2, \cdots, e_n\}$. For any e_i, there is a positive integer k_i, such that e_i occurs in the plenary power $e_i^{[k_i]}$ and e_i also occurs in $e_i^{[k_i+1]}$, since M_X is aperiodic. Let k_i be the least number that has this property. Now consider e_1, without loss of generality, we can assume that $k_1 = 1$, $\rho_1 L(e_1) \neq 0$,

$$L(e_1) = p_{11}e_1 + \sum_{k \in \Lambda_1} p_{k1}e_k, \quad p_{k1} \neq 0, \ k \in \Lambda_1,$$

where Λ_1 is not empty and $p_{11} \neq 0$. Otherwise, $\langle e_1 \rangle$ will be a proper subalgebra. From

$$L^2(e_1) = p_{11}^2 e_1 + p_{11}\sum_{i \in \Lambda_1} p_{i1}e_i + \sum_{i \in \Lambda_1} p_{i1}L(e_i),$$

we can see that once some e_i occurs in $L(e_1)$, it will keep in $L^n(e_1)$ for any power n. Since every e_j must occur in some plenary power of e_1, there is a positive integer m_1 so that $\{e_1, e_2, \cdots, e_n\} < L^{m_1}(e_1)$. Similarly, we have m_2 for e_2, $\cdots$, and m_n for e_n. Then, take $m_0 = Max\{m_1, m_2, \cdots, m_n\}$, we have

$$\rho_j L^{m_0}(e_i) \neq 0.$$

Therefore, L is positive.

Corollary 15. *The geometric multiplicity of eigenvalue 1 of the evolution operator of a simple aperiodic evolution algebra is 1.*

Theorem 30. *If M_X is a simple evolution algebra with period d, then the geometric multiplicity of eigenvalue 1 of the evolution operator is 1.*

Proof. By the decomposition Theorem 10 in Chapter 3, M_X can be written as

$$M_X = \Delta_0 \oplus \Delta_1 \oplus \cdots \oplus \Delta_{d-1}$$

and $L^d : \Delta_k \rightarrow \Delta_k$, $k = 0, 1, 2, \cdots, d-1$, and

$$L^d = l_0 \oplus l_1 \oplus \cdots \oplus l_{d-1},$$

where $l_k = L^d|_{\Delta_k}$, and it is positive (we give a proof of this claim below). If there are two vectors x, y, such that $L(x) = x$, $L(y) = y$, and $x \neq \lambda y$, then x has a unique decomposition according to the decomposition of M_X that is $x = x_0 + x_1 + \cdots + l_{d-1}$, and

$$L^d(x) = l_0(x_0) + l_1(x_1) + \cdots + l_{d-1}(x_{d-1})$$
$$= x_0 + x_1 + \cdots + x_{d-1}.$$

We get $l_k(x_k) = x_k$, since it is a direct sum. Similarly, $y = y_0 + y_1 + \cdots + y_{d-1}$ and $l_k(y_k) = y_k$, $k = 0, 1, \cdots, d-1$. Now, $x \neq \lambda y$, so there is an index k_0 so that $x_{k_0} \neq \lambda y_{k_0}$, but we know $l_{k_0}(x_{k_0}) = x_{k_0}$ and $l_{k_0}(y_{k_0}) = y_{k_0}$. This means that $L^d|_{\Delta_{k_0}} = l_{k_0}$ has two different eigenvectors for eigenvalue 1. This is a contradiction.

A proof of our claim that $L^d|_{\Delta_k}$ is positive:

Suppose $\Delta_k = Span\{e_{k,1}, e_{k,2}, \cdots, e_{k,t_k}\}$. Since d is the period, $\rho_{k,1}e_{k,1}^{[d]} \neq 0$, and there must be $e_{k,i}$ ($\neq e_{k,1}$) that occurs in $e_{k,1}^{[d]}$. Otherwise, Δ_k is the dimension of 1, which means d must be 1. So $L^d|_{\Delta_k}$ is positive. Therefore, we have that

$$l_k(e_{k,1}) = ae_{k,1} + be_{k,i} + \cdots,$$

then,

$$l_k^2(e_{k,1}) = a^2e_{k,1} + abe_{k,i} + bl_k(e_{k,i}) + \cdots.$$

We can see once $e_{k,i}$ occurs in $l_k(e_{k,1})$, $e_{k,i}$ will always keep in $l_k^n(e_{k,1})$ for any power n. Since every $e_{k,j}$ will occur in a certain $l_k^m(e_{k,1})$, there exists n_1 so that

$$\{e_{k,1}, e_{k,2}, \cdots, e_{k,t_k}\} < l_k^{n_1}(e_{k,1}).$$

Similarly, we have n_2 for $e_{k,2}, \cdots, n_{t_k}$ for e_{k,t_k}, so that

$$\{e_{k,1}, e_{k,2}, \cdots, e_{k,t_k}\} < l_k^{n_i}(e_{k,i}).$$

Set

$$m_k = \max\{n_1, n_2, \cdots, n_{t_k}\}.$$

For any $e_{k,i}$ and $e_{k,j}$

$$\rho_{k,j}l_k^{m_k}(e_{k,i}) = \rho_{k,j}(L^d|_{\Delta_k})^{m_k}(e_{k,i}) \neq 0.$$

Therefore, $l_k = L^d|_{\Delta_k}$ is positive.

Theorem 31. *Let M_X be a simple evolution algebra with period d, then the evolution operator has d eigenvalues that are the roots of unity. Each of them has an eigenspace of dimension one. And there are no other eigenvalues of modulus one.*

Proof. Since M_X is simple and periodical, it has a decomposition $M_X = \Delta_0 \oplus \Delta_1 \oplus \cdots \oplus \Delta_{d-1}$, and

$$L: \quad \Delta_k \to \Delta_{k+1}.$$

Denote $L|_{\Delta_k} = L_k$, then

$$L = L_0 + L_1 + \cdots + L_{d-1},$$
$$L^2 = L_1 L_0 + L_2 L_1 + \cdots + L_0 L_{d-1},$$
$$\cdots\cdots\cdots\cdots$$
$$L^d = L_{d-1} L_{d-2} \cdots L_1 L_0 \oplus L_0 L_{d-1} \cdots L_2 L_1 \oplus \cdots \oplus L_{d-1} \cdots L_0 L_{d-1}.$$

So, if denote

$$l_0 = L_{d-1} L_{d-2} \cdots L_1 L_0,$$
$$l_1 = L_0 L_{d-1} \cdots L_2 L_1,$$
$$\cdots\cdots,$$
$$l_{d-1} = L_{d-1} \cdots L_0 L_{d-1},$$

we have

$$L^d = l_0 \oplus l_1 \oplus \cdots \oplus l_{d-1},$$

and $l_k : \Delta_k \to \Delta_k$. If $L(x) = x$, then $L^d(x) = x$. x has a unique decomposition $x = x_0 + x_1 + \cdots + x_{d-1}$, so that

$$l_0(x_0) + l_1(x_1) + \cdots + l_{d-1}(x_{d-1}) = x_0 + x_1 + \cdots + x_{d-1}.$$

Therefore, $l_k(x_k) = x_k$, $k = 0, 1, 2, \cdots, d_{-1}$, which means that one is an eigenvalue of l_k(with geometric multiplicity 1 because l_k is positive). Thus, one is an eigenvalue of L^d, since L^d is a directed sum of l_k. Hence if λ is an eigenvalue of L, λ^d is an eigenvalue of L^d. So then $\lambda^d = 1$, or $\lambda_k = \exp \frac{2k\pi i}{d}$, $k = 0, 1, 2, \cdots, d - 1$, dth roots of unity are eigenvalues of L, which we prove as follows.

Now suppose that each λ_k is an eigenvalue of L, we prove it has geometric multiplicity 1. If $L(x) = \lambda_k x$, $L(y) = \lambda_k y$, $x \neq ky$, $x = x_0 + x_1 + \cdots + x_{d-1}$, and $y = y_0 + y_1 + \cdots + y_{d-1} \in \Delta_0 \oplus \Delta_1 \oplus \cdots \oplus \Delta_{d-1}$, then $L^d(x) = \lambda_k^d x = x$ and $L^d(y) = \lambda_k^d y = y$, so $l_k(x_k) = x_k$ and $l_k(y_k) = y_k$, $k = 0, 1, 2, \cdots, d - 1$. There is k_0, $x_{k_0} \neq k y_{k_0}$, but we have $l_{k_0}(x_{k_0}) = x_{k_0}$ and $l_{k_0}(y_{k_0}) = y_{k_0}$, which means that $l_{k_0} = L^d|_{\Delta_{k_0}}$ has two distinct eigenvectors, x_{k_0}, y_{k_0} for eigenvalue 1. But we know that positive operator l_k has an eigenspace of dimension 1 corresponding to eigenvalue 1. This contradiction means that the geometric multiplicity of each λ_k is one.

Each λ_k is really an eigenvalue of L, since each l_k is positive, $k = 0, 1, \cdots, d-1$, for their eigenvalue 1, let the corresponding eigenvectors are $y_0, y_1, \cdots, y_{d-1}$, respectively, $l_0(y_0) = y_0$, $l_1(y_1) = y_1, \cdots, l_{d-1}(y_{d-1}) = y_{d-1}$. Actually, $y_1 = L_0(y_0)$, $y_2 = L_1(y_1)$, $\cdots$, $y_{d-1} = L_{d-2}(y_{d-2})$, and $y_0 = L_{d-1}(y_{d-1})$ (up to a scalar). Remember $l_0 = L_{d-1}L_{d-2}\cdots L_1 L_0$, $l_1 = L_0 L_{d-1}\cdots L_2 L_1$, so $y_0 = L_{d-1}L_{d-2}\cdots L_1 L_0(y_0)$. Take the action of L_0 on both sides of the equation, we have $L_0(y_0) = L_0 L_{d-1}L_{d-2}\cdots L_1 L_0(y_0) = l_1 L_0(y_0)$. By the positivity of l_1, we have $y_1 = L_0(y_0)$. Similarly, we can obtain the other formulae. If we set $y = y_0 + y_1 + \cdots + y_{d-1}$, then $L(y) = y$, because

$$L(y) = L_0(y_0) + L_1(y_1) + \cdots + L_{d-1}(y_{d-1}) = y_0 + y_1 + \cdots + y_{d-1} + y_0 = y.$$

Now set

$$z_1 = y_0 + \lambda_1 y_1 + \lambda_2 y_2 + \cdots + \lambda_{d-1} y_{d-1} = \sum_{k=0}^{d-1} \lambda^k y_k,$$

$$\text{where } \lambda = \exp\frac{2\pi i}{d} \text{ and } \lambda_k = \lambda^k.$$

Then, we have

$$
\begin{aligned}
L(z_1) &= L(y_0) + \lambda_1 L(y_1) + \lambda_2 L(y_2) + \cdots + \lambda_{d-1} L(y_{d-1}) \\
&= L_0(y_0) + \lambda_1 L_1(y_1) + \cdots + \lambda_{d-1} L_{d-1}(y_{d-1}) \\
&= y_1 + \lambda_1 y_2 + \lambda_2 y_3 + \cdots + \lambda_{d-2} y_{d-1} + \lambda_{d-1} y_0 \\
&= \lambda_1^{-1}(\lambda_1 y_1 + \lambda_1^2 y_2 + +\lambda_1 \lambda_2 y_3 + \cdots + \lambda_1 \lambda_{d-2} y_{d-1} + \lambda_1 \lambda_{d-1} y_0) \\
&= \lambda_1^{-1}(y_0 + \lambda_1 y_1 + \lambda_2 y_2 + \lambda_3 y_3 + \cdots + \lambda_{d-1} y_{d-1}) \\
&= \lambda_{d-1} z_1,
\end{aligned}
$$

since $\lambda_1^{-1} = \lambda_{d-1}$. Set $z_2 = \sum_{k=0}^{d-1} \lambda^{2k} y_k$, then

$$
\begin{aligned}
L(z_2) &= \sum_{k=0}^{d-1} \lambda^{2k} L(y_k) = \sum_{k=0}^{d-1} \lambda^{2k} y_{k+1} = \lambda^{-2} \sum_{k=0}^{d-1} \lambda^{2(k+1)} y_{k+1} \\
&= \lambda_1^{-2} z_2 = \lambda_{d-2} z_2.
\end{aligned}
$$

Generally, set $z_k = \sum_{j=0}^{d-1} \lambda^{kj} y_k$, we have

$$L(z_k) = \lambda_{d-k} z_k.$$

And $z_{d-1} = \sum_{j=0}^{d-1} \lambda^{(d-1)j} y_j$, so we have $L(z_{d-1}) = \lambda_1 z_{d-1}$. Therefore, all λ_k are eigenvalues of L.

At last, we need to prove all eigenvalues of modulus one must be roots of dth unity. If $L(y) = \eta y$, $|\eta| = 1$, then $L^d(y) = \eta^d y$. y has a decomposition $y = y_0 + y_1 + \cdots + y_{d-1}$, and we have

$$L_0(y_0) + L_1(y_1) + \cdots + L_{d-1}(y_{d-1})$$
$$= \eta y_0 + \eta y_1 + \cdots + \eta y_{d-1},$$

then

$$L_0(y_0) = \eta y_1$$
$$L_1(y_1) = \eta y_2$$
$$\cdots\cdots\cdots\cdots$$
$$L_{d-1}(y_{d-1}) = \eta y_0.$$

Therefore, $L_1 L_0(y_0) = \eta^2 y_2$, $\cdots$, $L_{d-1}L_{d-2}\cdots L_1 L_0(y_0) = \eta^d y_0$. That is, $l_0(y_0) = \eta^d y_0$. Similarly, we can obtain $l_k(y_k) = \eta^d y_k$. Since each l_k is positive, then either $\eta^d = 1$ or $|\eta^d| < 1$. Because $|\eta| = 1$, we have $\eta^d = 1$, where η is a dth root of unity.

Corollary 16. *Let M_X be a finite dimensional evolution algebra, then any eigenvalue of its evolution operator of modulus one is a root of unity. The roots of dth unity are eigenvalues of L, if and only if M_X has a simple evolution subalgebra with period d.*

Proof. The first part of the corollary is obvious from the previous Theorem 31. If M_X has an evolution subalgebra with period d, as the proof of Theorem 31, the roots of dth unity are eigenvalues. Inversely, if L has an eigenvalue of root of dth unity, for example λ, $L(x) = \lambda x$, then we write x as a linear combination of basis $x = \sum_{i \in \Lambda_x} a_i e_i$, $i \in \Lambda_x$, $a_i \neq 0$, where Λ_x is a subset of the index set. Let $A_x = \langle e_i | i \in \Lambda_x \rangle$ be an evolution subalgebra generated by e_i, $i \in \Lambda_x$. Then A_x is a simple algebra with period d.

4.3.3 Spectrum of an evolution algebra at zeroth level

Theorem 32. *Let M_X be an evolution algebra of finite dimension, then the geometric multiplicity of the eigenvalue one of its evolution operator is equal to the number of simple evolution subalgebras of M_X.*

Proof. We know that the evolutionary operator L has a fixed point x_0. L, as a linear transformation of D, has eigenvalue 1 and an eigenvector with nonnegative components. Suppose that $M_X = A_1 \oplus \cdots \oplus A_n \,\overset{\bullet}{+}\, B_0$ is the decomposition of M_X, then

$$L: \quad A_k \cap D \to A_k \cap D, \quad k = 1, 2, \cdots, n$$

since $L(A_k) \subset A_k$. Since $A_k \cap M_0$ is still compact, Brouwer's fixed point theorem (Schauder theorem) can be applied to the restriction of L to get a fixed point in $A_k \cap M_0$, say x_k, $L(x_k) = x_k$, $k = 1, 2, \cdots, n$. Each x_k belongs to the eigenspace V_1 of eigenvalue 1. Since they do not share the same coordinate, $\{x_1, \cdots, x_n\}$ is an independent set. Thus $\dim V_1 \geq n$. On the other hand, for any vector $x \in V_1$, $x = \sum_{i=1}^{m} a_i e_i$ and $L(x) = x$. So $L^k(x) = x$ for any integer k. To finish the proof, we need the following statement.

Claim: If e_t is transient, then $\left\| \rho_t L^k(e_i) \right\| \to 0$ for any generator e_i, when $k \to \infty$.

Proof of the claim: Since $\sum_{k=1}^{\infty} \left\| \rho_t L^k(e_t) \right\| < \infty$, if e_t can not be accessible from e_i, $\left\| \rho_t L^k(e_i) \right\| = 0$ for any k. If e_t can be accessible from $e_i, \left\| \rho_t L^{k_0}(e_i) \right\| \neq 0$ for some k_0. Then $\sum_{k=1}^{\infty} \left\| \rho_t L^k(e_i) \right\| = \sum_{k=1}^{k_0} \left\| \rho_t L^k(e_i) \right\| + \sum_{k=k_0}^{\infty} \left\| \rho_t L^k(e_i) \right\| \leq c \sum_{k=1}^{\infty} \left\| \rho_t L^k(e_t) \right\| \leq \infty$, where c is a constant. Thus $\left\| \rho_t L^k(e_i) \right\| \to 0$.

Now, from this claim, we have $\left\| \rho_t L^k(x) \right\| \to 0$, when $k \to \infty$. Then we have $\rho_t(x) = \rho_t L^k(x) = 0$. This means that

$$x = \sum_{e_i \notin B_0} a_i e_i.$$

Therefore, we can rewrite x according to the decomposition $M_X = A_1 \oplus \cdots \oplus A_n \overset{\bullet}{+} B_0$, $x = y_1 + y_2 + \cdots + y_n$, $y_i \in A_i$. Since A_i is simple, y_i must be of the form of kx_i. Thus $\dim V_1 \leq n$. In a word, $\dim V_1 = n$.

We summarize here. Let M_X be an evolution algebra, we have a decomposition $M_X = A_1 \oplus A_2 \oplus \cdots \oplus A_n \overset{\bullet}{+} B_0$. Denote the period of A_k by d_k (d_k can be 1), then the evolution operator L has the following eigenvalues:

- 1 with the geometric multiplicity n;
- Roots of dth unity; each root d_k of dth unity has geometric multiplicity 1, $k = 0, 1, 2, \cdots, n$;
- In the zeroth transient space, the eigenvalue of the evolutionary operator is strictly less than 1.

4.4 Hierarchies of General Markov Chains and Beyond

4.4.1 Hierarchy of a general Markov chain

- Theorem of semi-direct-sum decomposition: Let M_X be a connected evolution algebra corresponding to Markov chain X. As a vector space, M_X has a decomposition

$$M_X = A_1 \oplus A_2 \oplus \cdots \oplus A_{n_0} \overset{\bullet}{+} B_0,$$

where A_i, $i = 1, 2, \cdots, n$, are all simple evolution subalgebras, $A_i \cap A_j = \{0\}$ for $i \neq j$, and B_0 is a subspace spanned by transient generators. We also call B_0 the 0th transient space of Markov chain X. Probabilistically, if the chain starts at some *0th* simple evolution subalgebra A_i, the chain will never leave the simple evolution subalgebra and it will run within this A_i forever. If it starts at the *0th* transient space B_0, it will eventually enter some *0th* simple subalgebra.

- The *1st* structure of X and the decomposition of B_0, as in Chapter 3, we have every first level concepts and the decomposition of B_0

$$B_0 = A_{1,1} \oplus A_{1,2} \oplus A_{1,3} \oplus \cdots \oplus A_{1,n_1} \overset{\bullet}{+} B_1$$

where $A_{1,i}$, $i = 1, 2, \cdots, n_1$, are all the first simple evolution subalgebras of B_0, $A_{1,i} \cap A_{1,j} = \{0\}, i \neq j$, and B_1 is the first transient space that is spanned by the first transient generators. When Markov chain X starts at the first transient space B_1, it will eventually enter a certain first simple evolution subalgebra $A_{1,j}$. Once the chain enters some first simple evolution subalgebra, it will sojourn there for a while and eventually go to some *0th* simple algebra.

- We can construct the *2nd* induced evolution algebra over the first transient space B_1, if B_1 is connected and can be decomposed. If the *kth* transient space B_k is disconnected, we will stop with a direct sum of reduced evolution subalgebras. Otherwise, we can continue to construct evolution subalgebras until we get a disconnected subalgebra. Generally, we can have a hierarchy as follows:

$$M_X = A_{0,1} \oplus A_{0,2} \oplus \cdots \oplus A_{0,n_0} \overset{\bullet}{+} B_0$$
$$B_0 = A_{1,1} \oplus A_{1,2} \oplus \cdots \oplus A_{1,n_1} \overset{\bullet}{+} B_1$$
$$B_1 = A_{2,1} \oplus A_{2,2} \oplus \cdots \oplus A_{2,n_2} \overset{\bullet}{+} B_2$$
$$\cdots\cdots\cdots\cdots\cdots\cdots\cdots\cdots\cdots\cdots\cdots$$
$$B_{m-1} = A_{m,1} \oplus A_{m,2} \oplus \cdots \oplus A_{m,n_m} \overset{\bullet}{+} B_m$$
$$B_m = B_{m,1} \oplus B_{m,2} \oplus \cdots \oplus B_{m,h},$$

where $A_{k,l}$ is the *kth* simple evolution subalgebra, $A_{k,l} \cap A_{k,l'} = \{0\}$ for $l \neq l'$, B_k is the *kth* transient space, and B_m can be decomposed as a direct sum of the *mth* simple evolution subalgebras. When Markov chain X starts at the *mth* transient space B_m, it will enter some *mth* simple evolution subalgebra $A_{m,j}$. Then, after a period of time, it will enter some *(m − 1)th* simple evolution subalgebra. The chain will continue until it enters certain *0th* simple evolution subalgebra $A_{0,i}$.

4.4.2 Structure at the 0th level in a hierarchy

Stability of evolution operators

Theorem 33. *For an evolution algebra M_X, $x \in D$, that is,*

$$x = \sum_{i \in \Lambda_x}^{n} x_i e_i, \quad \sum_{i \in \Lambda_x}^{n} x_i = 1, \ and \ 0 \leq x_i \leq 1,$$

the image of $L^m(e_i)$ will definitely go to the sum of simple evolution subalgebras of M_X, when m goes to the infinite. (the evolution of algebra M_X will be stabilized with probability 1 into a simple evolution subalgebra over time).

Proof. In the proof of Theorem 28 in Chapter 3, we got $\rho_t L^m(e_i) \to 0$ for the transient generator e_t, when $m \to \infty$. Thus $\|\rho_{B_0} L^m(e_i)\| \to 0$. Therefore, for any $x \in D$, $\|\rho_{B_0} L^m(x)\| \to 0$. This means $L^m(x)$ will go to a certain simple subalgebra as time m goes to the infinity.

Fundamental operators

Let M_X be an evolution algebra, B_0 be its *0th* transient space. The fundamental operator can be defined to be the projection of the evolution operator to the *0th* transient space B_0, i.e.,

$$L_{B_0} = \rho_{B_0} L,$$

ρ_{B_0} is the projection to B_0.

Theorem 34. *Let M_X be an evolution algebra. If M_X has a simple evolution subalgebra and a nontrivial transient space, then the difference $I - L_{B_0}$ has an inverse operator*

$$F = (I - L_{B_0})^{-1} = I + L_{B_0} + L_{B_0}^2 + \cdots.$$

Proof. In the Banach algebra $BL(M \to M)$, if the spectrum radius of L_{B_0} is strictly less than 1, then we can get this conclusion directly by using a result in Functional Analysis. So we need to check the spectrum radius of L_{B_0}.

Suppose λ is any eigenvalue of L_{B_0}, the corresponding eigenvector is v, then

$$L_{B_0}(v) = \lambda v, \qquad \forall\, m,$$

for any m, we still have

$$L_{B_0}^m(v) = \lambda^m v,$$

$$|\lambda^m| \cdot \|v\| = \left\|L_{B_0}^m(v)\right\| \le \|\rho_{B_0} L^m(v)\| \to 0,$$

as $m \to \infty$, we shall have $|\lambda| < 1$.

Corollary 17. *(Probabilistic version)* $\|\rho_j F(e_i)\|$ *is the expected number of times that the chain is in state e_j from e_i, when e_i, e_j are both in a transient space.*

Proof. Consider

$$F = I + L_{B_0} + L_{B_0}^2 + \cdots + L_{B_0}^m + \cdots,$$

so $\rho_j L_{B_0}^m(e_i) = a e_j$, which means the chain is in e_j in the mth step (if $a \ne 0$) with probability a. If we define a random variable $X^{(m)}$ that equals 1, if the chain is in e_j after m steps and equals to 0 otherwise, then

$$P\{X^{(m)} = 1\} = \|\rho_{B_0} L^m(e_i)\|,$$
$$P\{X^{(m)} = 0\} = 1 - \|\rho_{B_0} L^m(e_i)\|,$$
$$E(X^{(m)}) = P\{X^{(m)} = 1\} \cdot 1 + P\{X^{(m)} = 0\} \cdot 0 = \|\rho_{B_0} L^m(e_i)\|.$$

So, we have

$$E(X^{(0)} + X^{(1)} + \cdots + X^{(m)}) = \|\rho_{B_0} L^0(e_i)\| + \|\rho_{B_0} L(e_i)\| + \cdots + \|\rho_{B_0} L^m(e_i)\|$$
$$= \|\rho_{B_0} L^0(e_i) + \rho_{B_0} L(e_i) + \cdots + \rho_{B_0} L^m(e_i)\|.$$

When $m \to \infty$, we obtain

$$\|\rho_j F(e_i)\| = E \sum_{m=0}^{\infty} X^{(m)}.$$

Time to absorption

Definition 11. *Let e_i be a transient generator of an evolution algebra M_X. If there is an integer, such that $L_{B_0}^m(e_i) = 0$, we say e_i is absorbed in the mth step.*

Theorem 35. *Let $T(e_i)$ be the expected number of steps before e_i is absorbed from e_i. Then $T(e_i) = \|F(e_i)\|$.*

Proof. By Corollary 17, $\|\rho_j F(e_i)\|$ is the expected number of times that the chain is in state e_j from e_i (starting from e_i). So when we take sum over all the 0th transient space B_0, we will get the result

$$T(e_i) = \sum_{e_j \in B_0} \|\rho_j F(e_i)\| = \|F(e_i)\|.$$

As to the second equation, it is easy to prove, since F is the sum of any image of e_i under all powers of L_{B_0}.

Probabilities of absorption by *0th* simple subalgebras

Theorem 36. *Let $M_X = A_1 \oplus A_2 \oplus \cdots \oplus A_r \overset{\bullet}{+} B_0$ be the decomposition of M_X. If e_i is a transient generator, eventually it will be absorbed. The probability of absorption by a simple subalgebra A_k is given by $\|L_{A_k} F(e_i)\|$, where $L_{A_k} = \rho_{A_k} L$ is the projection to subalgebra A_k.*

Proof. We write $L_{A_k} F(e_i)$ out as follows

$$L_{A_k} F(e_i) = L_{A_k}(e_i) + L_{A_k} L_{B_0}(e_i) + L_{A_k} L_{B_0}^2(e_i) + \cdots \cdots.$$

We can see the coefficient of term $L_{A_k} L_{B_0}^2(e_i)$ is the probability that e_i is absorbed by A_k in the mth step. So when we take sum over times, we will obtain the total probability of absorption.

Remark 8.

$$\sum_{k=1}^{r} \|L_{A_k} F(e_i)\| = 1.$$

4.4.3 *1st* structure of a hierarchy

For an evolution algebra M_X, we have the *1st* structure

$$M_X = A_{0,1} \oplus A_{0,2} \oplus \cdots \oplus A_{0,n_0} \overset{\bullet}{+} B_0$$
$$B_0 = A_{1,1} \oplus A_{1,2} \oplus \cdots \oplus A_{1,n_1} \overset{\bullet}{+} B_1.$$

We define

$$L_1 = L_{B_1} = \rho_{B_1} L$$

to be the 1st fundamental operator.

Theorem 37. *Let M_X be an evolution algebra. If it has the 1st simple evolution subalgebra and the nontrivial 1st transient space, then the difference between the identity and the 1st fundamental operator, $I - L_1$, has an inverse operator, and*

$$F_1 = (I - L_1)^{-1} = I + L_1 + L_1^2 + \cdots.$$

Proof. The proof is easy, since the spectrum radius of L_1 is strictly less than 1.

Corollary 18. *$\|\rho_j F_1(e_i)\|$ is the expected number of times that the chain is in state e_j from e_i, where e_i and e_j are both in the 1st transient space.*

Proof. The proof is the same as that of Corollary 17.

Time to absorption at the *1st* level

Definition 12. *Let e_i be a 1st transient generator of an evolution algebra, i.e., $e_i \in B_1$. If there is an integer, such that $L_1^k(e_i) = 0$, we say that e_i is absorbed in the kth step at the 1st level.*

Theorem 38. *Let $T_1(e_i)$ be the expected number of steps before e_i is absorbed at the 1st level from e_i, $e_i \in B_1$, then $T_1(e_i) = \|F_1(e_i)\|$.*

Proof. The proof is the same as that of Theorem 35.

Probabilities of absorption by 1st simple subalgebras

Theorem 39. *Let $B_0 = A_{1,1} \oplus A_{1,2} \oplus \cdots \oplus A_{1,n_1} \overset{\bullet}{+} B_1$ be the decomposition of the 0th transient space of M_X. If $e_i \in B_1$, e_i will eventually be absorbed (leave space B_1). The probability of absorption by a simple 1st subalgebra $A_{1,k}$ is given by $\|L_{A_{1,k}} F_1(e_i)\|$, where $L_{A_{1,k}} = \rho_{A_{1,k}} L_{B_0}$ is the projection to the subalgebra $A_{1,k}$.*

Remark 9.

$$\sum_{k=1}^{n_1} \|L_{A_{1,k}} F_1(e_i)\| \leq 1.$$

4.4.4 *kth* structure of a hierarchy

Completely similarly, the *2nd* fundamental operator and other terms can be defined over the 1st structure of the hierarchy, and the corresponding theorems can be obtained. If an evolution algebra has N levels in the hierarchy, we can define the *(N−1)th* fundamental operator and other terms, we will also have the corresponding theorems.

Relationships between different levels in a hierarchy

Proposition 12. *For any generator $e_i \in A_{\delta,k}$, e_i will be in $A_{\zeta,l}$ with probability $\left\|L_{A_{\zeta,l}}F(e_i)\right\|$; the whole algebra $A_{\delta,k}$ will be in $A_{\zeta,l}$ with probability*

$$\frac{\left\|\sum_{e_i \in A_{\delta,k}} L_{A_{\zeta,l}}F(e_i)\right\|}{d(A_{\delta,k})},$$

where $d(A_{\delta,k})$ is the dimension of the δth subalgebra $A_{\delta,k}$, $0 \le \zeta < \delta$.

Proof. By the theorem of absorption probability, the first statement is just a repetition. For the second one we just need to sum the absorption probabilities over all the generators in the δth subalgebra $A_{\delta,k}$. Then normalizing this quantity by dividing the sum by the dimension of $A_{\delta,k}$, we shall get the probability that the whole algebra $A_{\delta,k}$ will be in $A_{\zeta,l}$.

The sojourn time during a simple evolution subalgebra

Suppose the evolution algebra M_X has a hierarchy as follows:

$$B_{m,1} \oplus B_{m,2} \oplus \cdots \oplus B_{m,h} = B_m$$

$$A_{m,1} \oplus A_{m,2} \oplus \cdots \oplus A_{m,n_m} \dotplus B_m = B_{m-1}$$

$$\cdots\cdots\cdots\cdots$$

$$A_{1,1} \oplus A_{1,2} \oplus \cdots \oplus A_{1,n_1} \dotplus B_1 = B_0$$

$$A_{0,1} \oplus A_{0,2} \oplus \cdots \oplus A_{0,n_0} \dotplus B_0 = M_X.$$

Then we have the following statements:

- We start at some head $B_{m,j}$ or a distribution v over B_m, the sojourn time during B_m (the expected number of steps or times before the chain leaves B_m) is given by
$$\left\|F_{B_m}(v)\right\|,$$
where $F_{B_m} = I_{B_m} + L_{B_m} + L_{B_m}^2 + \cdots = \sum_{k=0}^{\infty} L_{B_m}^k$.
- The sojourn time during $A_{m,1} \oplus A_{m,2} \oplus \cdots \oplus A_{m,n_m}$ is given by
$$\left\|F_{B_{m-1}}(v)\right\| - \left\|F_{B_m}(v)\right\|.$$

- The sojourn time during $A_{m,k}$, denoted by $m_{A_{m,k}}(v)$, is given by

$$m_{A_{m,k}}(v) = \left\| \rho_{A_{m,k}} F_{B_{m-1}}(v) \right\|.$$

- The sojourn time during $A_{k,1} \oplus \cdots \oplus A_{k,n_k}$, $k = 1, 2, \cdots, m$, is given by

$$\left\| F_{B_{k-1}}(v) \right\| - \left\| F_{B_k}(v) \right\|.$$

- Proposition (about sojourn times)

$$\sum_{k=1,l=1}^{m,n_k} m_{A_{k,l}}(v) + m_{B_m}(v) = \|F(e_i)\|.$$

Since the direction of chain moving along the hierarchy structure is limited from a higher indexed subalgebra to lower indexed ones, and it never goes back to higher indexed subalgebras if it once goes to a lower indexed subalgebra, so there is no overlap or uncover time to be considered before the chain enters some subalgebra in the *0th* level.

Example 4. If M_X has a decomposition as follows

$$M_X = A_0 \overset{\bullet}{+} B_0$$
$$B_0 = A_1 \overset{\bullet}{+} B_1$$
$$B_1 = A_2 \overset{\bullet}{+} B_2$$
$$\cdots \cdots$$
$$B_{m-1} = A_m \overset{\bullet}{+} B_m,$$

which satisfies $L(B_m) \subset A_m \cup B_m$, $L(A_m) \subset A_m \cup A_{m-1}$, $\cdots \cdots$, $L(A_1) \subset A_1 \cup A_0$, then we have

$$m_{A_k}(e_i) = m_{B_{k-1}}(e_i) - m_{B_k}(e_i), \quad k = 0, 1, \cdots, m,$$

where

$$m_{B_k}(e_i) = \|F_k(e_i)\| = \sum_{m=0}^{\infty} (\rho_{B_k} L)^m(e_i), \quad (F_0 = F).$$

Proof. We need to prove first

$$\rho_{A_1} F(e_i) = F_0(e_i) - F_1(e_i)$$
$$= \sum_{m=0}^{\infty} (\rho_{B_0} L)^m(e_i) - \sum_{m=0}^{\infty} (\rho_{B_1} L)^m(e_i)$$

by comparing them term by term. We look at

$$\rho_{B_0} L - \rho_{B_1} L = \rho_{A_1} \rho_{B_0} L,$$

this formula is true because $B_0 = A_1 \overset{\bullet}{+} B_1$. Let $\rho_{B_0} L(e_i) = u_1 + v_1$, $u_1 \in B_1$, $v_1 \in A_1$, we see,

$$
\begin{aligned}
(\rho_{B_0} L)^2 (e_i) &= (\rho_{B_0} L)(\rho_{B_0} L)(e_i) \\
&= (\rho_{B_0} L)(u_1 + v_1) = \rho_{B_0} L(u_1) + \rho_{B_0} L(v_1) \\
&= (\rho_{B_1} L)^2 (e_i) + \rho_{A_1} (\rho_{B_0} L)^2 (e_i)
\end{aligned}
$$

or

$$
\begin{aligned}
(\rho_{B_0} L)^2 &= (\rho_{A_1} L + \rho_{B_1} L)^2 \\
&= (\rho_{A_1} L)^2 + (\rho_{B_1} L)^2 + \rho_{A_1} L \rho_{B_1} L + \rho_{B_1} L \rho_{A_1} L \\
&= (\rho_{B_1} L)^2 + (\rho_{A_1} L)(\rho_{A_1} L + \rho_{B_1} L) \\
&= (\rho_{B_1} L)^2 + \rho_{A_1} L \rho_{B_0} L \\
&= (\rho_{B_1} L)^2 + \rho_{A_1} (\rho_{B_0} L)^2,
\end{aligned}
$$

since $\rho_{B_1} L \rho_{A_1} L = 0$. Thus,

$$
(\rho_{B_0} L)^2 (e_i) - (\rho_{B_1} L)^2 (e_i) = \rho_{A_1} (\rho_{B_0} L)^2 (e_i).
$$

Suppose

$$
(\rho_{B_0} L)^n = (\rho_{B_1} L)^n + \rho_{A_1} (\rho_{B_0} L)^n,
$$

then we check,

$$
\begin{aligned}
(\rho_{B_0} L)^{n+1} &= (\rho_{A_1} L + \rho_{B_1} L)(\rho_{B_0} L)^n \\
&= (\rho_{A_1} L + \rho_{B_1} L)[(\rho_{B_1} L)^n + \rho_{A_1} (\rho_{B_0} L)^n] \\
&= \rho_{A_1} L (\rho_{B_1} L)^n + \rho_{A_1} L \rho_{A_1} (\rho_{B_0} L)^n + \rho_{B_1} L (\rho_{B_1} L)^n \\
&\quad + \rho_{B_1} L \rho_{A_1} (\rho_{B_0} L)^n \\
&= (\rho_{B_1} L)^n + \rho_{A_1} L [(\rho_{B_1} L)^n + \rho_{A_1} (\rho_{B_0} L)^n] \\
&= (\rho_{B_1} L)^{n+1} + \rho_{A_1} (\rho_{B_0} L)^{n+1},
\end{aligned}
$$

by using $\rho_{B_1} L \rho_{A_1} (\rho_{B_0})^n = 0$ and $\rho_{A_1} \rho_{B_0} = \rho_{A_1}$. By induction, we finish the proof.

Remark 10. By this Example, we see that under a certain condition, the sojourn times can be computed step by step over the hierarchial structure of an evolution algebra.

4.4.5 Regular evolution algebras

Regular Markov chains are irreducible Markov chains. For a regular chain, it is possible to go from every state to every state after certain fixed number of steps. Their evolution algebras are simple and aperiodic. We may call these evolution algebras "regular evolution algebras." We will have a fundamental limit theorem for this type of algebras.

Definition 13. *Let A be a commutative algebra, we define semi-principal powers of a with b, a, b ∈ A, as follows:*

$$a * b = a \cdot b$$
$$a^2 * b = a \cdot (a \cdot b) = a \cdot (a * b)$$
$$a^3 * b = a \cdot [a \cdot (a \cdot b)] = a \cdot (a^2 * b)$$
$$\cdots\cdots$$
$$a^n * b = a \cdot (a^{n-1} * b).$$

Theorem 40. *Let M_X be a regular evolution algebra with a generator set $\{e_1\ e_2\ \cdots\ e_r\}$, $x = \sum_{i=1}^{r} \alpha_i e_i$ be any probability vector; that is, $0 < \alpha_i < 1$ and $\sum_{i}^{r} \alpha_i = 1$. Then,*

$$limit_{n \to \infty} \theta^n * x = \sum_{i=1}^{r} \pi_i e_i,$$

where $\theta = \sum_{i=1}^{r} e_i$, and $\pi = \sum_{i=1}^{r} \pi_i e_i$ with $0 < \pi_i < 1$ and $\sum_{i}^{r} \pi_i = 1$, is constant probability vector.

Recall that for an evolution algebra the universal element θ has the same function as the evolution operator L does. Let us first prove a lemma related to positive evolution operators and then prove this theorem.

Lemma 12. *Let θ be the element corresponding to a positive evolution operator L and $c = Min\{\|\rho_i e_k^2\|,\ i, k \in \Lambda\}$. Let $y = \sum_{i=1}^{r} y_i e_i$, and $M_0 = Max\{\|\rho_i y\|,\ i \in \Lambda\}$, and $m_0 = Min\{\|\rho_i y\|,\ i \in \Lambda\}$. Let $M_1 = Max\{\|\rho_i \theta y\|,\ i \in \Lambda\}$ and $m_1 = Min\{\|\rho_i \theta y\|,\ i \in \Lambda\}$ for the element θy. Then*

$$M_1 - m_1 \le (1 - 2c)(M_0 - m_0).$$

Proof. Note that each coefficient of θy is a weighted average of the coefficients of y. The biggest possible weight would be $cm_0 + (1 - c)M_0$, and the smallest possible weighted average be $cM_0 + (1 - c)m_0$. Thus, $M_1 - m_1 \le (cm_0 + (1 - c)M_0) - (cM_0 + (1 - c)m_0)$; this is, $M_1 - m_1 \le (1 - 2c)(M_0 - m_0)$.

Let us give a brief proof of Theorem 40. Denote $M_n = Max\{\rho_i \theta^n * y, i \in \Lambda\}$ and $m_n = Min\{\rho_i \theta^n * y, i \in \Lambda\}$. Since each component of $\theta^n * y$ is an average of the components of $\theta^{n-1} * y$, we have $M_0 \ge M_1 \ge M_2 \ge \cdots$ and $m_0 \le m_1 \le m_2 \le \cdots$. Each sequence is monotone and bounded, $m_0 \le m_n \le M_n \le M_0$. Therefore, they have limits as n tends to infinity. If M is the limit of M_n and m the limit of m_n, $M - m = 0$. This can be seen from $M_n - m_n \le (1 - 2c)^n (M_0 - m_0)$, since $c < \frac{1}{2}$.

The Theorem 40 has an interesting consequence, and it is written as the following proposition.

Proposition 13. *Within a regular evolution algebra, the algebraic equation*

$$\theta \cdot x = x$$

has solutions, and the solutions form an one-dimensional linear subspace.

Now we provide statements relating to the mean first occurrence time.

Definition 14. *Let M_X be a simple evolution algebra with the generator set $\{e_1 \ e_2 \ \cdots \ e_n\}$, for any e_i, the expected number of times that e_i visits e_j for the first time is called the mean first occurrence time (passage time or visiting time), denote it by m_{ij}. Then by the definition*

$$m_{ij} = \sum_{m=1}^{\infty} m \left\| V_j^{(m)}(e_i) \right\|,$$

where $V_j^{(m)}$ is the operator of the first visiting to e_j at the mth step.

Remark 11. Since we work on simple evolution algebras, so

$$D_j(e_i) = \sum_{m=1}^{\infty} V_j^{(m)}(e_i) = e_j.$$

This definition makes sense.

Proposition 14. *Let M_X be a simple evolution algebra, we define*

$$F_j = \sum_{m=0}^{\infty} (\rho_j^0 L)^m.$$

Then we have

$$m_{ij} = \|F_j(e_i)\|, \quad if \quad i \neq j \,,$$
$$m_{ij} = r_{ij}, \quad if \quad i = j, \ the \ mean \ recurrence \ time.$$

Proof. Take $\rho_j^0 L = \rho_{e_j}^0 L$ as a fundamental operator, we have

$$\sum_{m=0}^{\infty} (\rho_j^0 L)^m = (I - \rho_j^0 L)^{-1}.$$

Taking derivative with respect to L as L is a real variable, and we have

$$\sum_{m=0}^{\infty} m(\rho_j^0 L)^{m-1} = (I - \rho_j^0 L)^{-2}.$$

Multiply by $\rho_j L$ from the left-hand side, we obtain

$$\sum_{m=0}^{\infty} m\rho_j L(\rho_j^0 L)^{m-1} = \rho_j L(I - \rho_j^0 L)^{-2}.$$

Then, when $i \neq j$,

$$\sum_{m=0}^{\infty} m\rho_j L(\rho_j^0 L)^{m-1}(e_i) = \rho_j L(I - \rho_j^0 L)^{-2}(e_i).$$

We have,

$$\rho_j L(I - \rho_j^0 L)^{-2}(e_i) = \rho_j L(I - \rho_j^0 L)^{-1}(I - \rho_j^0 L)^{-1}(e_i)$$

$$= \sum_{m=0}^{\infty} \rho_j L(\rho_j^0 L)^{m-1}(I - \rho_j^0 L)^{-1}(e_i)$$

$$= D_j(I - \rho_j^0 L)^{-1}(e_i) = D_j F_j(e_i).$$

Therefore,

$$m_{ij} = \sum_{m=0}^{\infty} m \left\| \rho_j L(\rho_j^0 L)^{m-1} \right\|$$

$$= \sum_{m=1}^{\infty} m \left\| V_j^{(m)}(e_i) \right\|$$

$$= \| D_j F_j(e_i) \| = \| F_j(e_i) \|.$$

When $i = j$,

$$r_j = \sum_{m=1}^{\infty} m \left\| V_j^{(m)}(e_i) \right\|,$$

r_j is the expected return time.

4.4.6 Reduced structure of evolution algebra M_X

As we know, by the reducibility of an evolution algebra, a simple evolution subalgebra can be reduced to an one-dimensional subalgebra. Now for the evolution algebra M_X corresponding to a Markov chain X, each simple evolution subalgebra can be viewed as one "big" state, since it corresponds to a "closed subset" of the state space. Then the following formulae give probabilities that higher indexed subalgebras move to lower indexed subalgebras.

- Moving from $B_{m,j}$ to $A_{k,l}$, $k = 0, 1, \cdots, m-1$, l can be any number that matches the chosen index k, with probability

$$\frac{1}{d(B_{m,j})} \sum_{e_i \in B_m} L_{A_{k,l}}(e_i),$$

where $d(B_{m,j})$ is the dimension of the evolution subalgebra $B_{m,j}$.

- Moving from $A_{k,l}$ to $A_{k',l'}$, $k' < k$, $k = 1, \cdots, m$, with probability

$$\frac{1}{d(A_{k,l})} \sum_{e_j \in A_{k,l}} L_{A_{k',l'}}(e_i).$$

4.4.7 Examples and applications

In this section, we discuss several examples to show algebraic versions of Markov chains, evolution algebras, also have advantages in computation of Markov processes. Once we use the universal element θ instead of the evolution operator in calculation, any probabilistic computation becomes an algebraic computation. For simple examples, we can deal with hands; for complicated examples, we just need to perform a Mathematica program for nonassociative setting symbolic computation. More advantages of evolution algebraic computation shall be revealed when a Markov chain has many levels in its hierarchy.

Example 5. A man is playing two slot-machines. The first machine pays off with probability p, the second with probability q. If he loses, he plays the same machine again; if he wins, he switches to the other machine. Let e_i be the state of playing the ith machine. We will form an algebra for this playing. The defining relations of the evolution algebra are

$$e_1 \cdot e_2 = 0,$$
$$e_1^2 = (1 - p)e_1 + pe_2,$$
$$e_2^2 = qe_1 + (1 - q)e_2.$$

The evolution operator is given by $\theta = e_1 + e_2$. If the man starts at a general state $\beta = a_1e_1 + a_2e_2$, the status after n plays is given by $\theta^n * \beta$. That is

$$(\theta \cdots \theta(\theta(\theta\beta)) \cdots).$$

Since $\theta\beta = (e_1 + e_2)(a_1e_1 + a_2e_2) = (a_1 + a_2q - a_1p)e_1 + (a_2 + a_1p - a_2q)e_2$, we can compute the semi-principal power and have

$$\theta^n * \beta = \frac{a_1p(1 - p - q)^n + a_1q + a_2q - a_2(1 - p - q)^n q}{p + q}e_1$$
$$+ \frac{a_1p + a_2p - a_1p(1 - p - q)^n + a_2(1 - p - q)^n q}{p + q}e_2.$$

It is easy to see that after infinite many times of plays, the man will reach the status $\frac{q}{p+q}e_1 + \frac{p}{p+q}e_2$. If $p = 1$ and $q = 1$, we have a cyclic algebra. That is $(e_i^2)^2 = e_i$. If $p = 0$ and $q = 0$, we have a nonzero trivial algebra. If one of these two parameters is zero, say $q = 0$, the algebra has one subalgebra and one transient space. Since $\theta \cdot e_2 = e_2$ in this case, the evolution operator can be represented by ρ_1e_1, and we have

$$F(e_1) = \sum_{n=0}^{\infty}(\rho_1 e_1)^n * e_1 = e_1 + (1-p)e_1 + (1-p)^2 e_1 + \cdots = \frac{1}{p}e_1.$$

So, the expected number that this man plays machine 1 is $\frac{1}{p}$.

Example 6. We continue the example 5. Let us suppose there are five machines available for this man to play. Playing the machine 1, he wins with probability p; if he loses, he play the machine 1 again, otherwise move to the machine 2. Playing the machine 2, he wins with probability q; if he loses, he play the machine 2 again, otherwise move to the machine 3. Playing the machine 3, he loses with probability $1 - r - s$, wins with probability $r + s$; when he wins, he moves to the machine 2 with probability r and move to the machine 4 with probability s. Once he plays machine 4 and 5, he cannot move to other machines. The machine 4 pays off with probability u, the machine 5 with probability v; if he loses, he play the same machine again.

As the example 5, the defining relations are given by

$$e_1^2 = (1-p)e_1 + pe_2, \quad e_2^2 = (1-q)e_2 + qe_3,$$
$$e_3^2 = re_2 + (1-r-s)e_3 + se_4, \quad e_i \cdot e_j = 0,$$
$$e_4^2 = (1-u)e_4 + ue_5, \quad e_5^2 = ve_4 + (1-v)e_5.$$

The algebra has a decomposition $M(X) = A_0 \dot{+} B_0$, and $B_0 = A_1 + B_1$, where $A_0 = \langle e_4, e_5 \rangle$, which is a subalgebra; $B_0 = \text{Span}(e_1, e_2, e_3)$, which is the 0th transient space; $A_1 = \langle e_2, e_3 \rangle_1$, which is a 1st subalgebra, and $B_1 = \text{Span}(e_1) = Re_1$, which is the first transient space. We ask what are the expected numbers that this man plays the same machine when he starts at the machine 1, 2, and 3, respectively. From the algebraic structure of this evolution algebra, we can decompose the evolution operator L or correspondingly decompose $\theta = \sum_{i=1}^{5} e_i$ as $\theta_1 = e_1$, $\theta_2 = e_2 + e_3$, and $\theta_3 = e_4 + e_5$. Starting at the machine 1, it is easy to compute that

$$e_1 + \theta_1 * e_1 + \theta_1^2 * e_1 + \theta_1^3 * e_1 + \cdots = \frac{1}{p}e_1.$$

That gives us the mean number he plays the machine, which is $\frac{1}{p}$. Generally, we need to compute $\sum_{k=0}^{\infty}(\theta_1 + \theta_2)^k * e_1$. We perform a Mathematica program to compute it, or compute it by hands inductively. We get the result which is $\frac{1}{p}e_1 + \frac{r+s}{qs}e_2 + \frac{1}{s}e_3$. So, when this man starts to play the machine 1, the mean number of playing the machine is $\frac{1}{p}$, the mean number of playing the machine 2 is $\frac{r+s}{qs}$ and the mean number of playing the machine 3 is $\frac{1}{s}$. Starting at the machine 2, we need to compute

$$e_2 + \theta_2 * e_2 + \theta_2^2 * e_2 + \theta_2^3 * e_2 + \cdots.$$

We perform a Mathematica program to compute this nonassociative summation, it gives us $\frac{r+s}{qs}e_2 + \frac{1}{s}e_3$. (It also can be obtained inductively.) Thus, the expected number that this man plays the machine 2 is $\frac{r+s}{qs}$, when he start at the machine 2; and the expected number he plays the machine 3 is $\frac{1}{s}$. Similarly, we can get the expected number that he plays the machine 3 is $\frac{r}{qs}$. Once he moves to the machine 4 or 5, he will stay there for ever. As example 5, from a long run, he will play the machine 4 with probability $\frac{v}{u+v}$, play the machine 5 with probability $\frac{u}{u+v}$.

Example 7. We modify an example from Kempthorne [42] as our example of applications to Mendelian genetics, a simple case of Wright-Fisher models. In the next chapter, we will apply evolution algebras to Non-Mendelian genetics. Here we consider the simplest case, where only two genes are involved in each generation, a and A. Hence any individual must be of gene type aa or aA or AA. Assume A dominates a, then AA is a pure dominant, aA is a hybrid, and aa is a pure recessive individual. Then a pair of parents must be of one of the following six types: (AA, AA), (aa, aa), (AA, Aa), (aa, Aa), (AA, aa), (Aa, Aa). We think of each pair of parents as one self-reproduction animal with four genes. The offspring is produced randomly. In its production, it is s times as likely to produce a given animal unlike itself than a given animal like itself. Thus s measures how strongly "opposites attract each other." We take into account that in a simple dominance situation, AA and Aa type animal are alike as far as appearance are concerned. We set $(AA, AA) = e_1$, $(aa, aa) = e_2$, $(AA, Aa) = e_3$, $(aa, Aa) = e_4$, $(AA, aa) = e_5$, and $(Aa, Aa) = e_6$. Then, we have an algebra generated by these generators and subject to the following defining relations:

$$e_1^2 = e_1, \ e_2^2 = e_2, \ e_i \cdot e_j = 0,$$

$$e_3^2 = \frac{1}{4}e_1 + \frac{1}{2}e_3 + \frac{1}{4}e_6, \ e_5^2 = e_6,$$

$$e_4^2 = \frac{1}{2(s+1)}e_2 + \frac{s}{s+1}e_4 + \frac{1}{2(s+1)}e_6,$$

$$e_6^2 = \frac{1}{4(s+3)}e_1 + \frac{1}{4(3s+1)}e_2 + \frac{1}{s+1}e_3 + \frac{2s(s+1)}{(s+3)(3s+1)}e_4$$
$$+ \frac{s(s+1)}{(s+3)(3s+1)}e_5 + \frac{1}{s+1}e_6.$$

We see that there are two subalgebras generated by e_1 and e_2, respectively, which correspond to pure strains: pure dominant and pure recessive; the transient space B_0 is spanned by the rest generators. Now we ask the following questions: when a hybrid parent starts to reproduce, what's the mean generations to reach a pure strain? How do the parameter s affect these quantities? To answer these questions, we need to compute $F(e_i) = \sum_{k=0}^{\infty} (\rho_{B_0} \theta)^k * e_i$ for each hybrid parent e_i. We perform a Mathematica program, and get

$$F(e_3) = \frac{4(s^2 + 5s + 2)}{2s^2 + 7s + 3}e_3 + \frac{2s(s+1)^2}{2s^2 + 7s + 3}e_4 + \frac{s^2 + s}{2s^2 + 7s + 3}e_5 + \frac{3s^2 + 10s + 3}{2s^2 + 7s + 3}e_6,$$

$$F(e_4) = \frac{6s+2}{2s^2+7s+3}e_3 + \frac{4s^3+13s^2+12s+3}{2s^2+7s+3}e_4 + \frac{s^2+s}{2s^2+7s+3}e_5 + \frac{3s^2+10s+3}{2s^2+7s+3}e_6,$$

$$F(e_5) = \frac{12s+4}{2s^2+7s+3}e_3 + \frac{4s(s+1)^2}{2s^2+7s+3}e_4 + \frac{4s^2+9s+3}{2s^2+7s+3}e_5 + \frac{6s^2+20s+6}{2s^2+7s+3}e_6,$$

$$F(e_6) = \frac{12s+4}{2s^2+7s+3}e_3 + \frac{4s(s+1)^2}{2s^2+7s+3}e_4 + \frac{2s^2+2s}{2s^2+7s+3}e_5 + \frac{6s^2+20s+6}{2s^2+7s+3}e_6.$$

From the theory developed in this chapter, the value

$$\|F(e_3)\| = \frac{2s^3+12s^2+33s+11}{2s^2+7s+3}$$

is the mean generations that when the parent (AA, Aa) starts to produce randomly, the genetic process reaches the pure strains. Similarly,

$$\|F(e_4)\| = \frac{4s^3+17s^2+29s+8}{2s^2+7s+3},$$

$$\|F(e_5)\| = \frac{4s^3+18s^2+45s+13}{2s^2+7s+3},$$

$$\|F(e_6)\| = \frac{4s^3+16s^2+38s+10}{2s^2+7s+3}$$

are the mean generations that when parents (aa, Aa), (AA, aa), and (Aa, Aa) start to produce randomly, the genetic processes reach the pure strains, respectively. We see that all these mean generations are increasing functions of the parameter s. Therefore, large s has the effect of producing more mixed offsprings. It is expected that a large s would slow down the genetic process to a pure strain.

5

Evolution Algebras and Non-Mendelian Genetics

In this chapter, we shall apply evolution algebra theory to non-Mendelian genetics. In the first section, we give a brief reflection of how non-Mendelian genetics motivated the development evolution algebras. In section 2, we review the basic biological components of non-Mendelian genetics and the inheritance of organelle genes; we also give a general algebraic formulation of non-Mendelian genetics. In section 3, we use evolution algebras to study the heteroplasmy and homoplasmy of organelle populations, and show that concepts of algebraic transiency and algebraic persistency relate to biological transitory and stability, respectively. Coexistence of triplasmy in tissues of patients with sporadic mitochondrial disorders is studied as well. In section 4, we apply evolution algebra theory to the study of asexual progenies of *Phytophthora infestans*, an important agricultural pathogen.

5.1 History of General Genetic Algebras

There is a long history of recognizing algebraic structures and properties in Mendelian genetics. Mendel first exploited some symbols [30], which is quite algebraically suggestive to express his genetic laws. In fact, it was later termed "Mendelian algebras" by several authors. In the 1920s and 1930s, general genetic algebras were introduced. Serebrowsky [31] was the first to give an algebraic interpretation of the sign "×," which indicated sexual reproduction, and to give a mathematical formulation of Mendel's laws. Glivenkov [32] continued to work at this direction and introduced the so-called Mendelian algebras for diploid populations with one locus or two unlinked loci. Independently, Kostitzin [33] also introduced a "symbolic multiplication" to express Mendel's laws. The systematic study of algebras occurring in genetics was due to I. M. H. Etherington. In his series of papers [34], he succeeded in giving a precise mathematical formulation of Mendel's laws in terms of nonassociative algebras. He pointed out that the nilpotent property is essential to these genetic algebras and formulated it in his definitions of train algebras and baric algebras. He also introduced the concept of commutative duplication by which the gametic algebra of a randomly mating population is associated with a zygotic algebra. Besides Etherington, fundamental contributions have been made by Gonshor [35], Schafer [36], Holgate [37,38], Hench [39], Reiser [40], Abraham [41], Lyubich [47], and Worz-Busekos [46]. It is worth mentioning two unpublished work in the field. One is Claude Shannon's Ph.D thesis submitted in 1940 (MIT) [43]. Shannon developed an algebraic method to predict the genetic makeup in future generations of a population starting with arbitrary frequencies. Particularly, the results for genetic algebras with three loci was quite interesting. The other one is Charles Cotterman's Ph.D thesis that was also submitted in 1940 (the Ohio State University) [44] [45]. Cotterman developed a similar system as Shannon did. He also put forward a concept of derivative genes, now called "identical by descent." During the early days in this area, it appeared that the general genetic algebras or broadly defined genetic algebras (by these term we mean any algebra that has been used in Mendelian genetics) can be developed into a field of independent mathematical interest, because these algebras are in general not associative and do not belong to any of the well-known classes of nonassociative algebras, such as Lie algebras, alternative algebras, or Jordan algebras. They possess some distinguished properties that lead to many interesting mathematical results. For example, baric algebras, which have nontrivial representations over the underlying field, and train algebras, whose coefficients of rank equations are only functions of the images under these representations, are new subjects for mathematicians. Until the 1980s, the most comprehensive reference in this area was Worz-Busekos' book [46]. More recent results, such as evolution theory in genetic algebras, can be found in Lyubich's book [47]. A good survey article is Reed's paper [48].

General genetic algebras are the product of interactions between biology and mathematics. Mendelian genetics offers a new subject to mathematics: general genetic algebras. The study of these algebras reveals the algebraic structures of Mendelian genetics, which always simplifies and shortens the way to understand genetic and evolutionary phenomena. Indeed, it is the interplay between the purely mathematical structures and the corresponding genetic properties that makes this area so fascinating. However, after Baur [49] and Correns [50] first detected that chloroplast inheritance departed from Mendel's rules, and much later, mitochondrial gene inheritance were also identified in the same way, non-Mendelian inheritance of organelle genes became manifest with two features – uniparental inheritance and vegetative segregation. Non-Mendelian genetics is now a basic language of molecular geneticists. Logically, we can ask what new subject non-Mendelian genetics offers to mathematics, and what mathematics offers to understanding of non-Mendelian genetics. It is clear that non-Mendelian genetics introduces new mathematical challenges. When we try to formulate non-Mendelian genetics as algebras, we at least need a new idea to formulate reproduction in non-Mendelian genetics as multiplication in algebras. Actually, "evolution algebras" [24] stems from this new idea.

5.2 Non-Mendelian Genetics and Its Algebraic Formulation

5.2.1 Some terms in population genetics

Before we discuss the mathematics of genetics, we need to acquaint ourselves with the necessary language from biology. DNA is a polymer and consists of a long chain of monomers called **nucleotides**. The DNA molecule is said to be a **polynucleotide**. Each nucleotide has three parts: a sugar, a nitrogen containing ring-structure called a **base**, and a phosphate group. DNA molecules have a very distinct and characteristic three-dimensional structure known as the double helix. It is the sequence of the bases in the DNA polynucleotide that encodes the genetic information. A **gene** is a unit of information and corresponds to a discrete segment of DNA that encodes the amino acid sequence of a polypeptide. In higher organisms, the genes are present on a series of extremely long DNA molecules called **chromosomes**. For example, in humans there are estimated 50–100,000 genes arranged on 23 chromosomes. Organisms with a double set of chromosomes are called **diploid organisms**. For example, humans are diploid. Organisms with one set of chromosomes are called haploid organisms. For instant, most fungi and a few algae are **haploid organisms**. The different variants of a gene are referred to as **alleles**. Biologists refer to individuals with two identical copies of a gene as being **homozygous**; and individuals with two different copies of the same gene as being **heterozygous**. Reproduction of organisms can take place by asexual or sexual processes. **Asexual reproduction** involves the production of a new individual(s) from cells or tissues of a preexisting organism. This process is common in plants and in many microorganisms. It can involve simple binary fission in unicellular microbes or the production of specialized asexual spores. Asexual reproduction allows some genetic changes in offspring by chance. **Sexual reproduction** differs, in that it involves fusion of cells (gametes) derived from each parent, to form a zygote. The genetic processes involved in the production of gametes also allow for some genetic changes from generation to generation. Sexual reproduction is limited to species that are diploid or have a period of their life cycle in the diploid state. The division of somatic cells is called **mitosis**; and the division of meiotic cells is called **meiosis**. **Prokaryote chromosomes** consist of a single DNA, which is usually circular, with only a small amount of associated protein. **Eukaryotes** have several linear chromosomes, and the DNA is tightly associated with large amounts of protein.

5.2.2 Mendelian vs. non-Mendelian genetics

Although most of heredity of nuclear genes obeys Mendel's laws, the inheritance of organelle is not Mendelian. Before we introduce the basic of organelle biology, we need review basic knowledge of Mendelian and non-Mendelian genetics.

Following Birky's paper [51], there are five aspects in comparison of Mendelian genetics and non-Mendelian genetics:

(1) During asexual reproduction, alleles of nuclear genes do not segregate: heterozygous cells produce heterozygous daughters. This is because all chromosomes in nuclear genomes are replicated once and only once in interphase and mitosis ensures that both daughter cells get one copy of each chromosome. In contrast, alleles of organelle genes in heteroplasmic cells segregate during mitotic as well as meiotic divisions to produce homoplasmic cells. This is because in the vegetative division of the organelles, some copies of the organelle genome can replicate more than others by chance or in response to selective pressures or intrinsic advantages in replication, and alleles can segregate by chance.

(2) Alleles of a nuclear gene always segregate during meiosis, with half of the gametes receiving one allele and half the other. Alleles of organelle genes may or may not segregate during meiosis; the mechanisms are the same as for vegetative segregation.

(3) Inheritance of nuclear genes is biparental. Organelle genes are often inherited from only one parent, uniparental inheritance.

(4) Alleles of different nuclear genes segregate independently. Organelle genes are nearly always on a single chromosome and recombination is often severely limited by uniparental inheritance or failure of organelles to fuse and exchange genomes.

(5) Fertilization is random with respect to the genotype of the gametes. This is the only part of Mendel's model that applies to organelle as well as nuclear genes.

We now review the basic of organelle biology.

Cell organelles include chloroplasts and mitochondria, which are substructural units within cells. **Chloroplasts** and **mitochondria** of eukaryotes contain their own DNA genomes. These DNA genomes vary considerably in size but are usually circular. They probably represent primitive prokaryote organisms that were incorporated into early eukaryotes and have coevolved in a **symbiotic relationship**. The organelles have their own ribosomes and synthesize some of their own proteins, but others are encoded by nuclear genes. When all of the mitochondria DNA (mtDNA) within each cell becomes genetically homogeneous, we have **homoplasmic cells**; and when mutant mtDNA molecules coexist with original mtDNA, we have **heteroplasmic cells**. Evolutionarily, chloroplasts and mitochondria have **endosymbiotic origin**. They have evolved from free-living prokaryotes. They are now integral parts of eukaryotic cells retaining only vestiges of their original genomes. Yet the genes encoded in these organelles are vital to their function as are the ones they have shed into the nucleus over the millennium. Bio-energetically, chloroplasts and mitochondria complement one another. Chloroplasts derive energy from light that is employed for splitting water and the production of molecular oxygen. The electrons produced from the splitting of water are used via the photo-synthetic electron transport chain to drive photosynthetic phosphorylation. Ultimately, molecular CO_2 is reduced by the protons and electrons derived from water and is converted into carbohydrates by the soluble enzymes of the chloroplast stroma. The mitochondrion, in contrast, catalyze the aerobic oxidation of reduced carbon compounds via soluble enzymes of the tricarboxylic acid cycle found in its matrix. The electrons produced by the oxidation of reduced carbon compounds flow via the respiratory electron transport chain and drive oxidative phosphorylation. The electrons and protons derived from the oxidation of reduced carbon compounds convert molecular oxygen to water and CO_2 is released as an oxidation product of the tricarboxylic acid cycle. In summary, the chloroplast reduces CO_2 and splits water with the release of CO_2, while the mitochondrion oxidizes reduced carbon compounds with the formation of CO_2 and water. However, chloroplasts and mitochondria are not simple energy-generating and utilizing systems. A vast array of other metabolic processes goes on within their confines as well, which are just as much key to the health and well-being of the cell as electron transport and energy generation. Genetically, mitochondrial and chloroplast (extra-nuclear) genomes are self-replicating units (but not physiologically) independent of the nuclear genome. Remarkably, the best way to think about chloroplast and mitochondrial gene inheritance is in terms of populations of organelle genes inside a single cell or cell line, subject to mutation, selection, and random drift. Chloroplasts vary in size, shape, and number per cell. A typical flowering plant has 10–200 chloroplasts. All animal cells contain many copies of mitochondrial genomes, on the order of thousands of molecules of mtDNA [52]. Therefore, it is appropriate to treat the group of chloroplasts or mitochondria in a cell as a population. This way we can take a perspective of population genetics and utilize methods in population genetics to study organelle inheritance. This is **intracellular population genetics of organelles**.

Vegetative segregation is the most general characteristics of the inheritance of organelle genes, occurring in both mitochondria and chloroplasts in all individuals or clones of all eukaryotes. In other words, **uniparental inheritance** is a major means of genetic transmission. More knowledge will be introduced when we construct various evolution algebras in the next section.

5.2.3 Algebraic formulation of non-Mendelian genetics

Let us consider a population of organelles in a cell or a cell clone, and suppose that there are n different genotypes in this organelle population. Denote these genotypes by $g_1, g_2, \ldots, g_n$. According to the point (3) in Subsection 5.2.2, the crossing of genotypes is impossible since it is uniparental inheritance. Mathematically, we set

$$g_i \cdot g_j = 0,$$

for $i \neq j$. According to the point (2) in Subsection 5.2.2, alleles of organelle genes may or may not segregate during meiosis following vegetative segregation, so the frequency of each gene in the next generation could be variant. According to the point (4) in Subsection 5.2.2, intramolecular and intermolecular recombination within a lineage provides evidence that one organelle genotype could produce other different genotypes. Therefore, we can mathematically define,

$$g_i^2 = \sum_{i=1}^{n} \alpha_{ij} g_j,$$

where α_{ij} is positive number that can be interpreted as the rate of genotype g_j produced by genotype g_i. Now, we have the algebra defined by generators $g_1, g_2, \ldots, g_n$, which are subject to these relations.

Obviously, this is a very general definition. But it is general enough to include all non-Mendelian inheritance phenomena. As an example, we will look at organelle heredity in the next section.

5.3 Algebras of Organelle Population Genetics

5.3.1 Heteroplasmy and homoplasmy

Organelle population geneticists are usually concerned about a special case where there are two different phenotypes or genotypes: homoplasmic and heteroplasmic. Let us denote the heteroplasmic cell by g_0, and the two different type of homoplasmic cells by g_1 and g_2, respectively. Just suppose g_1 and g_2 are mutant and wild-type, respectively. From the inheritance of organelles we know that heteroplasmic parents can produce both heteroplasmic progeny and homoplasmic progeny, and homoplasmic parents can only produce homoplasmic progeny with the same type where mutation is not considered for the moment. Figure 5.1 shows the Wright-Fisher model for organelle genes.

Therefore, we have the following reproductive relations.

$$g_0^2 = \pi g_0 + \alpha g_1 + \beta g_2, \tag{5.1}$$
$$g_1^2 = g_1, \tag{5.2}$$
$$g_2^2 = g_2; \tag{5.3}$$

and for $i \neq j$, $i, j = 0, 1, 2$,

$$g_i \cdot g_j = 0; \tag{5.4}$$

where π, α, β are all positive real numbers. Actually, these numbers can be taken as the segregation rates of corresponding types. For any specific

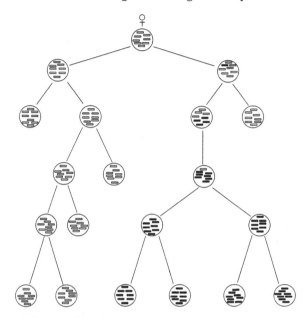

Fig. 5.1. Wright-Fisher model for organelle genes

example, we can determine these coefficients by combinatorics or modified Wright-Fisher model.

Thus, we have an evolution algebra, denoted by A_h, generated by g_0, g_1, and g_2 and subject to the above defining relations (5.1)–(5.4).

By our knowledge of evolution algebras, algebraic generator g_0 is transient; g_1 and g_2 are persistent. Because g_1 and g_2 do not communicate, we have two simple subalgebras of A_h generated by g_1 and g_2, respectively. Biologically, g_0 is transitory as N. W. Gillham pointed out [53]; g_1 and g_2 are of stable homoplasmic cell states. By transitory, biologists mean that the cells of transitory are not stable; they are just transient phases, and they will disappear eventually after certain cell generations. This property is imitated by algebraic transiency. By biological stability, we mean it is not changeable over time, and it is kept the same from generation to generation. This property is imitated by algebraic persistency.

The puzzling feature of organelle heredity is that the heteroplasmic cells eventually disappear and the homoplasmic progenies are observed. The underlying biological mechanisms are still unknown. Actually, it is a intensive research field currently, since it is related to aging and many other diseases caused by mitochondrial mutations [54], [55]. However, by the theory of evolution algebras we could mathematically understand this phenomenon. Because g_0 is transient, g_1 and g_2 are persistent, by evolution algebra theory we can eventually have two simple subalgebras of A_h. These two subalgebras are of zero-th in the hierarchy of this evolution algebra, and thus they are stable.

The subalgebra generated by g_1 is homoplasmic and mutant; the subalgebra generated by g_2 is homoplasmic and wild-type. Moreover, the mean time T_h to reach these homoplasmic progeny is given by

$$T_h = \frac{1}{1 - \pi}.$$

If we now consider a mutant to be lost, say gene g_2 will be lost, we have the following several ways to model this phenomenon. The algebraic generator set is still $\{g_0, g_1, g_2\}$.

First, we think that g_2 disappears in a dramatic way, that is

$$g_2^2 = 0.$$

Other defining relations are (5.1), (5.2), and (5.4). Thus, the evolution algebra we get here is different from A_h. It has one nontrivial simple subalgebra that is corresponding to homoplasmic progeny generated by g_1.

Second, we consider that g_2 gradually mutates back to g_1, that is

$$g_2^2 = \eta g_1 + \rho g_2,$$

where η is not zero and could be 1. And other defining relations are (5.1), (5.2), and (5.4). Although we eventually have one simple subalgebra by these relations, the evolution path is different.

Third, we consider that g_2 always keeps heteroplasmic property, that is

$$g_2^2 = \eta g_0 + \rho g_2.$$

Other defining relations are still (5.1), (5.2), and (5.4). Eventually, we have homoplasmic progenies that are all g_1. That is the only simple subalgebra generated by g_1.

In conclusion, we have four different evolution algebras derived from the study of homoplasmy. They are not the same in skeletons. Therefore, their dynamics, which are actually genetic evolution processes, are different. However, it seems that we need to look for the biological evidences for defining these different algebras. In Ling et al. [55], several hypothetical mechanisms were put forward for the establishment of homoplasmy. These hypothetical mechanisms are actually corresponding to four different algebraic structures above.

5.3.2 Coexistence of triplasmy

In mitochondrial genetics, if we consider different genotypes of mutants instead of just two different phenotypes of homoplasmy and heteroplasmy, we will have higher dimensional algebras that contain more genetic information. Recently, in Tang et al. [56], it studied the dynamical relationship among wild-type and rearranged mtDNAs.

Large-scale rearrangements of human mitochondrial DNA (including partial duplications and deletion) are found to be associated with a number of human disorders, including Kearns-Sayre syndrome, progressive external ophthalmoplegia, Pearson's syndrome, and some sporadic myopathies. Each patient usually harbors a heteroplasmic population of wild-type mitochondrial genomes (wt-mtDNA) together with a population of a specific partially deleted genome (Δ-mtDNA) in clinically affected tissues. These patients also harbor a third mtDNA species, a partial duplication (dup-mtDNA), as well. To study the dynamic relationship among these genotypes, authors of paper [56] cultured cell lines from two patients. After a long-term (6 months, 210–240 cell divisions) culture of homoplasmic dup-mtDNAs from one patient, they found the culture contained about 80% dup-mtDNA, 10% wt-mtDNA, and 10% Δ-mtDNA. After a long-term culture of the heteroplasmic that contains wt-mtDNA and Δ-mtDNA from the same patient, they did not find any new cell species, although there were fluctuations of percentages of these two cell populations. From this same patient, after culturing Δ-mtDNA cell line for two years, they did not find any new cell species. Now, let us formulate this genetic dynamics as an algebra.

Denote triplasmic cell population by g_0 that contain dup-mtDNA, wt-mtDNA, and Δ-mtDNA, denote heteroplasmy that contains dup-mtDNA and wt-mtDNA by g_1, heteroplasmy that contains dup-mtDNA and Δ-mtDNA by g_2, heteroplasmy that contains wt-mtDNA and Δ-mtDNA by g_3, and homoplasmy dup-mtDNA by g_4, homoplasmy wt-mtDNA by g_5, homoplasmy Δ-mtDNA by g_6. According to the genetic dynamical relations described earlier, we set algebraic defining relations as follows:

$$g_0^2 = \beta_{00}g_0 + \beta_{01}g_1 + \beta_{02}g_2 + \beta_{03}g_3,$$
$$g_1^2 = \beta_{14}g_4 + \beta_{15}g_5,$$
$$g_2^2 = \beta_{24}g_4 + \beta_{26}g_6,$$
$$g_3^2 = \beta_{35}g_5 + \beta_{36}g_6,$$
$$g_4^2 = \beta_{44}g_4 + \beta_{45}g_5 + \beta_{46}g_6,$$
$$g_5^2 = \beta_{54}g_4 + \beta_{56}g_6,$$
$$g_6^2 = \beta_{64}g_4 + \beta_{65}g_5,$$

and for $i \neq j$, $i, j = 0, 1, \ldots, 6$,

$$g_i \cdot g_j = 0.$$

And the generator set is $\{g_0, g_1, \ldots, g_6\}$. This algebra has three levels of hierarchy. On the *0th* level, it has one simple subalgebra generated by g_4, g_5, and g_6. These three generators are algebraic persistent. Biologically, they consist of the genotypes that can be observed, and genetically stable. On the *1st* level, it has three subalgebras; each of them is of dimension 1. On the *2nd* level, there is one subalgebra generated by g_0. Generators on the *1st* and *2nd* levels are all algebraic transient. They are unobservable biologically.

If we have more information about the reproduction rates β_{ij}, we could quantitatively compute certain relevant quantities. For example, let us set

$$\beta_{00} = \beta_{01} = \beta_{02} = \beta_{03} = \frac{1}{4},$$

$$\beta_{14} = \beta_{15} = \frac{1}{2},$$

$$\beta_{24} = \beta_{26} = \frac{1}{2},$$

$$\beta_{35} = \beta_{36} = \frac{1}{2},$$

$$\beta_{44} = \frac{5}{6},$$

$$\beta_{45} = \beta_{46} = \frac{1}{12},$$

$$\beta_{54} = \frac{2}{3}, \beta_{56} = \frac{1}{3},$$

$$\beta_{64} = \frac{2}{3}, \beta_{65} = \frac{1}{3}.$$

Then we can compute the long-term frequencies of each genotype in the culture. Actually, the limit of the evolution operator will give the answer. Suppose we start with a transient genotype g_0, then we have a starting vector $v_0 = (1, 0, \ldots, 0)'$. As time goes to infinity, we have

$$\lim_{n \to \infty} L^n v_0 = (0, \ldots, 0, 0.80, 0.10, 0.10)'.$$

Therefore, to this patient, we can see the algebraic structure of his mitochondrial genetic dynamics. Besides the experimental results we could reproduce by our algebraic model, we could predict that there are several transient phases. These transient phases are algebraic transient generators of the algebra. They are important for medical treatments. If we could have drugs to stop the transitions during the transient phases of mitochondrial mutations, we could help these disorder patients.

5.4 Algebraic Structures of Asexual Progenies of *Phytophthora infestans*

In this section, we shall apply evolution algebra theory to the study of algebraic structures of asexual progenies of *Phytophthora infestans* based on experimental results in Fry and Goodwin [57]. The basic biology of *Phytophthora infestans* and related experiments are first briefly introduced. Then we will construct evolution algebras for each race of *Phytophthora infestans*. Most of our biological materials is taken from Fry and Goodwin [57] and [58].

5.4.1 Basic biology of *Phytophthora infestans*

Oomycetes are a group of organisms in a kingdom separated from the true fungi, plants, or animals. They are included in the Kingdom Protoctista or Chromista. This group of organisms is characterized by the absence of chitin in the cell walls (true fungi contain chitin), zoospores with heterokont flagella (one whiplash, one tinsel) borne in sporangia, diploid nuclei in vegetative cells, and sexual reproduction via antheridia and oogonia [58]. The genus *Phytophthora* contains some species including *P. infestans* that are heterothallic (A1 and A2 mating types) and some that are homothallic. The Chromista organism *P. infestans* (Mont.) *de Bary*, the cause of potato and tomato late blight, is the most important foliar and tuber pathogen of potato worldwide. The Irish Potato Famine is a well-known result of these early epidemics. Tomato late blight was detected sometime later and has also been a persistent problem. Most scientists agree that the center of origin of *P. infestans* is in the highlands of central Mexico and that this region has been the ultimate source for all known migrations. It was the only location where both mating types of *P. infestans* were found prior to the 1980s. Outside Mexico, *P. infestans* populations were dominated by a single clonal lineage that are confined to asexual reproduction [59]. Sexual reproduction of *P. infestans*, associated with genetic recombination during meiosis in the antheridium or the oogonium, is a major mechanism of genetic variation in this diploid organism. However, other mechanism of genetic variability may have a significant role in creating new variants of this pathogen. Mutation, mitotic recombination, and parasexual recombination are the most common mechanism of genetic variability in the absence of sexual reproduction [60]. The most important aspect of genetic variability in plant pathogens is the variability in pathogenicity and virulence toward the host. Virulence variability in *P. infestans* populations is recognized as a major reason for failure of race specific genes for resistance in cultivated potato management strategy. The **race** concept as applied to *P. infestans* refers to possession of certain virulence factors. Isolates sharing the same virulence factors are considered to be a race that can be distinguished from other races possessing other groups of virulence factors. Characterization of isolates to different races is based on their interaction with major genes for resistance in potato. So far 11 major genes for resistance have been identified in Solanum [61].

In paper [57], five parental isolates of *P. infestans*, PI-105, PI-191, PI-52, PI-126, and PI-1, collected from Minnesota and North Dakota in 1994–1996, were chosen to represent different race structures. Single zoospore progenies were generated from each of the parental strains by inducing asexual zoospore production. The proportion of zoospores that developed into vegetative colonies varied from 2 to 50% depending on the parental isolate.

The parental isolate PI-1 produced very small zoospores and the percent recovery of colonies was very low. Other parental isolates produced large-sized zoospores and showed higher levels of developed colonies. In total, 102 single zoospore isolates were recovered, 20 isolates from isolate PI-105, 29 isolates from PI-191, 28 isolates from PI-52, 14 isolates from PI-126, and 11 isolates from PI-1. These single zoospore demonstrated different levels of variability for virulence. Although some single zoospore isolates showed the same virulence as their parental isolate, others showed lower or higher virulence than the isolate from which they were derived. Single zoospore isolates derived from PI-1 (11 isolates) were identical in virulence to their parental isolate. Single zoospore isolates derived from isolate PI-191 (29 isolates) showed low levels of variability for virulence compared with their parental isolate; 73% of these isolates (21 isolates) retained the same virulence pattern as their parent. Four isolates gained additional virulence to R8 and R9. One isolate had additional virulence to R9, which was stable. The other two showed lower virulence compared with the parental isolate. Six races were identified from the single zoospore isolates of the parental isolate PI-191.

Single zoospore isolates derived from isolate PI-126 showed higher levels of variability for virulence. Three isolates in this series gained virulence to both R8 and R9, three isolates gained additional virulence to R8, six isolates gained additional virulence to R9, and only two isolates retained the same virulence level of the parental isolate. Four races were identified within this series of isolates.

Isolates derived from the parental isolate PI-52 were highly variable for virulence. The overall trend in this series of isolates was toward lower virulence relative to the parental isolate. The total number of races identified from this parental isolate is 12.

The single zoospore progeny isolates derived from isolate PI-105 were highly variable for virulence. In this series of isolates, there was a tendency for reduced virulence of the single zoospore isolates compared with their parent. Thirteen races were identified from this set of isolates.

5.4.2 Algebras of progenies of *Phytophthora infestans*

To mathematically understand the complexity of structure of progenies of *P. infestans*, we assume that there are 11 loci in genome of *P. infestans* corresponding to the resistant genes, or 11 phenotypes corresponding to the resistant genes, denote by $\{c_1, c_2, \ldots, c_{11}\}$, and if c_j functions (is expressed), the progeny resists gene R_j. Any nonrepeated combination of these c_j could form a race mathematically. So, we can have 2048 races. For simplicity, we just record a virulence part of a race by E_i, the complement part is avirulence. For example, $E_i = \{c_2, c_3, c_5, c_8, c_{10}\}$ represents race type $c_2 c_3 c_5 c_8 c_{10} / c_1 c_4 c_6 c_7 c_9 c_{11}$. Take these 2048 races as generators set, we then have a free algebra over the real number field R. Since reproduction of zoospore progeny is asexual, the generating relations among races are types of evolution algebras. That is,

$$E_i^2 = \sum p_{ij} E_j,$$

and if $i \neq j$

$$E_i \cdot E_j = 0,$$

where p_{ij} are nonnegative numbers. If we interpret p_{ij} as frequency, we have $\sum p_{ij} = 1$. If we have enough biological information about the generating relations among the races or within one race, we could write the detailed algebraic relations.

For example, let us look at the race PI-126P and its progenies. PI-126P has race type $E_1 = \{c_1, c_2, c_3, c_4, c_5, c_6, c_7, c_{10}, c_{11}\}$. It has four different type of progenies:

$$\{c_1, c_2, c_3, c_4, c_5, c_6, c_7, c_8, c_{10}, c_{11}\} = E_2,$$
$$\{c_1, c_2, c_3, c_4, c_5, c_6, c_7, c_9, c_{10}, c_{11}\} = E_3,$$
$$\{c_1, c_2, c_3, c_4, c_5, c_6, c_7, c_8, c_9, c_{10}, c_{11}\} = E_4,$$

and E_1 itself. Actually, these four types of progenies are biologically stable, and we could eventually observe them as outcomes of asexual reproduction. These four types of progenies, as generators algebraically, are persistent elements. There could have been many transient generators that produce biologically unstable progenies. These unstable progenies serve as intermediate transient generations, and produces stable progenies. A simple evolution algebra without intermediate transient generations that we could construct for race PI-126P may have the following defining relations:

$$E_1^2 = p_1 E_2 + q_1 E_3,$$
$$E_2^2 = p_2 E_1 + q_2 E_4,$$
$$E_3^2 = p_3 E_1 + q_3 E_4,$$
$$E_4^2 = r_1 E_1 + r_0 E_4;$$

and if $i \neq j$,

$$E_i \cdot E_j = 0.$$

If we know the frequency p_j of the jth race in the population as in paper [57], we could easily set the above coefficients. For example, suppose all coefficients have the same value, 0.5, then the algebra generated by PI-126P is a simple evolution algebra. Biologically, this simple evolution algebra means that each race can reproduce other races within the population. We can also compute that the period of each generator, for each race, is 2. This means to reproduce any race itself at least needs two generations. Eventually, frequencies of races E_1, E_2, E_3, and E_4 in the population are $\frac{1}{3}, \frac{1}{6}, \frac{1}{6}$, and $\frac{1}{3}$ respectively. This can be done by computing

$$\lim_n L^n(E_1),$$

where L is the evolution operator of the simple algebra.

Now, let us assume that there exists an intermediate transient generation, therefore there exists a transient race, E_5, in the developing process of progeny population of PI-126P. We just assume that E_5 is $\{c_1, c_2, c_3, c_4, c_5, c_6, c_7, c_{10}, \}$. Usually, it is very difficult to observe the transient generation biologically. Our evolution algebra is now generated by E_1, E_2, E_3, E_4, and E_5. The defining relations we choose are given

$$E_1^2 = p_1 E_2 + q_1 E_3,$$
$$E_2^2 = p_2 E_1 + q_2 E_4 + r_2 E_5,$$
$$E_3^2 = p_3 E_1 + q_3 E_4,$$
$$E_4^2 = r_1 E_1 + r_0 E_4,$$
$$E_5^2 = 0$$

and if $i \neq j$,

$$E_i \cdot E_j = 0.$$

We can verify that this evolution algebra has a simple subalgebra, which is just constructed above. We also claim that intermediate transient races will extinct, and they are not biologically stable. Mathematically, these intermediate transient races are nilpotent elements.

The progeny population of PI-52P shows a distinct algebraic feature.

There are 12 races in the progeny population of PI-52P, and the parental race is not in the population. We name these races as follows. According to paper [57]: $E_0 = \{c_3, c_4, c_7, c_8, c_{10}, c_{11}\}$, which is parental race, and the progenies are:

$$E_1 = \{c_3, c_7, c_{10}, c_{11}\},$$
$$E_2 = \{c_{10}, c_{11}\},$$
$$E_3 = \{c_1, c_3, c_7, c_{10}, c_{11}\},$$
$$E_4 = \{c_3, c_{10}, c_{11}\},$$
$$E_5 = \{c_1, c_2, c_3, c_{10}, c_{11}\},$$
$$E_6 = \{c_2, c_4, c_{10}, c_{11}\},$$
$$E_7 = \{c_1, c_{10}, c_{11}\},$$
$$E_8 = \{c_7, c_{11}\},$$
$$E_9 = \{c_7, c_{10}, c_{11}\},$$
$$E_{10} = \{c_3, c_4, c_7, c_{10}, c_{11}\},$$
$$E_{11} = \{c_1, c_3, c_4, c_7, c_{10}, c_{11}\},$$
$$E_{12} = \{c_2, c_3, c_4, c_{10}, c_{11}\}.$$

Thus, our evolution algebra is generated by E_0, E_1, ..., E_{12}. As to the defining relations, we need the detailed biological information, such as the frequency of each race in progeny population. However, E_0 must be a transient generator, an intermediate transient race in the progeny population, while all other generators must be persistent generators, biologically stable races that can be observed in experiments. For illustration, we give the defining relations below:

$$E_0^2 = \sum_{i=1}^{12} \frac{1}{12} E_i,$$

$$E_1^2 = \frac{1}{2} E_1 + \frac{1}{2} E_2,$$

for $2 \leq j \leq 11$,

$$E_j^2 = \frac{1}{3} E_{j-1} + \frac{1}{3} E_j + \frac{1}{3} E_{j+1},$$

and for $j = 12$,

$$E_{12}^2 = \frac{1}{2} E_{11} + \frac{1}{2} E_{12};$$

and if $i \neq j$,

$$E_i \cdot E_j = 0.$$

This algebra is not simple. But it has a simple subalgebra generated by $\{E_1, E_2, \ldots, E_{12}\}$. We know that this subalgebra forms a progeny population of parental race PI-52P. This subalgebra is aperiodic, which means biologically each race in progeny population could reproduce itself in the next generation. By computing

$$\lim_n L^n(E_0),$$

we get that in the progeny population, frequency of parental race E_0 is 0, frequencies of races E_1 and E_{12} both are 5.88%, frequencies of races E_2, E_3, ..., E_{11} all are 8.82%. This is the asymptotic behavior of the evolution operator.

Now let us add some intermediate transient races, biological unstable races, into the population. Suppose we have two such races, E_α and E_β. Theoretically, there are many ways to build an evolution algebra with these two transient generators based on the above algebra with biology information. Each way will carry different biological evolution information. Here, let us choose the following way to construct our evolution algebra.

The generator set is $\{E_\alpha, E_\beta, E_0, E_1, \ldots, E_{12}\}$. The set of defining relations is taken as

$$E_0^2 = pE_\alpha + qE_\beta,$$

$$E_\alpha^2 = \sum_{i=1}^{12} \frac{1}{12} E_i,$$

$$E_\beta^2 = \sum_{i=1}^{12} \frac{1}{12} E_i,$$

$$E_1^2 = \frac{1}{2} E_1 + \frac{1}{2} E_2,$$

for $2 \le j \le 11$,

$$E_j^2 = \frac{1}{3} E_{j-1} + \frac{1}{3} E_j + \frac{1}{3} E_{j+1},$$

and for $j = 12$

$$E_{12}^2 = \frac{1}{2} E_{11} + \frac{1}{2} E_{12};$$

and if $i \ne j$,

$$E_i \cdot E_j = 0.$$

Although this new algebra is not simple, it has a simple subalgebra that forms progeny population. Two unstable races, mathematically not necessarily nilpotent, will eventually disappear through producing other races. Whatever the values of p and q are, we eventually get the same frequency of each race in the population as that in the simple algebra above, except that E_α and E_β both have 0 frequency.

There is a trivial simple algebra generated by race PI-1P. If we denote PI-1P by E_{-1}, the progeny population is generated by E_{-1} which is subject to $E_{-1}^2 = E_{-1}$.

In paper [57], there are five different parental races in Minnesota and North Dakota from 1994 to 1996. If we want to study the whole structure of *P. infestans* population in Minnesota and North Dakota, we need to construct a big algebra that is reproduced by 5 parental races, PI-105P, PI-191P, PI-52P, PI-126P, and PI-1P. This algebra will have five simple subalgebras, which corresponds to the progeny subpopulations produced by five parental races. We also need to compute the frequency of each progeny subpopulation. This way, we encode the complexity of structure of progenies of *P. infestans* into an algebra.

Let us summarize what evolution algebras can provide to plant pathologists theoretically.

(1) Evolution algebra theory can predict the existence of intermediate transient races. Intermediate transient races correspond to algebraic transient elements. They are biologically unstable, and will extinct or disappear by producing other races after a period of time. If we can catch the intermediate transient races that do not extinct but disappear through producing other new races, and remove or kill them, we will easily stop the spread of late blight disease.

(2) Evolution algebra theory says that biologically stable races correspond to algebraic persistent elements. It predicts the periodicity of reproduction of stable races. This is helpful to understand the speed of spread of plant diseases.

(3) Evolution algebra theory can rerecover the existence of progeny subpopulation. Furthermore, because these progeny subpopulations correspond to simple subalgebras, each race in the same subpopulation shares the same dynamics of reproduction and spreading. Evolution algebras are, therefore, helpful to simplify the complexity of progeny population structure.

(4) Evolution algebra theory provides a way to compute the frequency of each race in progeny population given reproduction rates, which are algebra structural constants. Practically, these frequencies can be measured, and therefore reproduction rates could be computed by formulae in evolution algebras. Therefore, evolution algebras will be a helpful tool to study many aspects of asexual reproduction process, like that of Oomycetes, Phytophthora.

6

Further Results and Research Topics

In the final chapter, we list some further related results that we have obtained. Because of the limitation of time and space, we do not give the detailed proofs for most of the results, although some explanations or brief proofs are given. To promote further study and better understanding of the significance of evolution algebras, we also put forward some interesting open problems and related research topics.

6.1 Beginning of Evolution Algebras and Graph Theory

Definition 15. *Let* $G = (V, E)$ *be a graph,* V *be the set of vertices of* G, E *be the set of edges of* G. *We define an algebra as follows: taking* $V = \{e_1, e_2, \cdots, e_r\}$ *as the generator set and*

$$R = \left\{ e_i^2 = \sum_{e_k \in \Gamma(e_i)} e_k; \; e_i \cdot e_j = 0, \; i \neq j; i, j = 1, \; 2, \; \cdots, \; r \right\}$$

as the set of defining relations, where $\Gamma(e_i)$ *is the set of neighbors of* e_i. *Then the evolution algebra determined by this graph is a quotient algebra*

$$A(G) = \langle V \mid R \rangle$$
$$= \left\langle e_1, e_2, \cdots, e_r \mid e_i^2 - \sum_{e_k \in \Gamma(e_i)} e_k; \; e_i \cdot e_j, i \neq j; i, j = 1, \; 2, \; \cdots, \; r \right\rangle$$

Theorem 41. *If graphs* G_1 *and* G_2 *are isomorphic as graphs, then* $A(G_1)$ *and* $A(G_2)$ *are also isomorphic as algebras.*

Proof. Denote $G_1 = (V_1, E_1)$, $G_2 = (V_2, E_2)$, and an isomorphism from G_1 to G_2 by Φ. Suppose $V_1 = \{e_1, e_2, \cdots, e_n\}$. Then

$$\Phi(V_1) = V_2 = \{\Phi(e_1), \Phi(e_2), \cdots, \Phi(e_n)\},$$

and
$$\Phi(e_i) \sim \Phi(e_j) \text{ (neighborhood) if and only if } e_i \sim e_j.$$

Look at algebras $A(G_1)$ and $A(G_2)$, we extend the map Φ linearly

$$\overline{\Phi} : A(G_1) \rightarrow A(G_2)$$
$$\overline{\Phi}(x) = \sum_i a_i \Phi(e_i),$$

if $x = \sum_i a_i e_i$. Then $\overline{\Phi}$ will be an algebraic map:

$$e_i \cdot e_j = 0, \ i \neq j, \ \text{then} \ \Phi(e_i) \cdot \Phi(e_j) = 0.$$

For $\overline{\Phi}$, we see

$$\begin{aligned}
\overline{\Phi}(e_i \cdot e_j) &= \overline{\Phi}(0) = 0 \\
&= \Phi(e_i)\Phi(e_j) \\
&= \overline{\Phi}(e_i)\overline{\Phi}(e_j).
\end{aligned}$$

We also have

$$\overline{\Phi}(e_i^2) = \sum_{e_k \in \Gamma(e_i)} \Phi(e_k)$$
$$= \Phi(e_i)^2 = \overline{\Phi}(e_i)^2.$$

Therefore, $\overline{\Phi}$ is an algebraic map. $\overline{\Phi}$ is $1-1$ and onto. Thus, $A(G_1)$ and $A(G_2)$ are algebraically isomorphic.

Definition 16. *A commutative nonassociative algebra A is called graphicable if it has a set of generators $V = \{x_1, x_2, \cdots, x_n\}$ with the defining relations*

$$x_i^2 = \sum_{e_k \in V_i} x_k; \ x_i \cdot x_j = 0, \ i \neq j; \ i, j = 1, \ 2, \ \cdots, \ n,$$

where V_i is a subset of V.

Now, if A is a graphicable algebra, then we can associate a graph G to A as follows: the set of vertices of G is V; for each x_i, assign all vertices in V_i as its neighbors. Denote $G(A)$ as the graph defined by graphicable algebra A.

Definition 17. *Presentable isomorphic: Let A_1 and A_2 be two graphicable algebras with the same generator set. If they are algebraically isomorphic, we say A_1 and A_2 are presentable isomorphic.*

Theorem 42. *Let A_1 and A_2 be two graphicable algebras. If A_1 and A_2 are presentable isomorphic, then their associated graphs $G(A_1)$ and $G(A_2)$ are isomorphic.*

Proof. By the definition of the associated graph, $G(A_1)$ and $G(A_2)$ have the same vertex set, denoting it by $\{e_1, e_2, \cdots, e_r\}$. If the isomorphic map is Φ, then $\{e_1, e_2, \cdots, e_r\} = \{\Phi(e_1), \Phi(e_2), \cdots, \Phi(e_r)\}$. Since an isomorphic map preserves the algebraic relations, then

$$\Phi(e_i^2) = \Phi(e_i) \cdot \Phi(e_i) = \sum_{k \in \Lambda_i} \Phi(e_k),$$

where Λ_i is a subset of index set, determined by e_i, $i = 1, 2, \cdots, r$. Therefore, the incidence relations among associated graphs are preserved. Thus $G(A_1)$ and $G(A_2)$ are isomorphic.

Theorem 43. *Let G be a graph with the vertex set $V(G) = \{x_1, x_2, \cdots, x_r\}$, L be the evolution operator of graphicable algebra $A(G)$, and suppose*

$$L^n(e_i) = n_{i1}x_1 + n_{i2}x_2 + \cdots + n_{ir}x_r,$$

then n_{ij} is the total number of paths with length n from vertex e_i to vertex e_j. If $n_{ij} = 0$, this means there is no path with length n from vertex e_i to vertex e_j.

Destination operators

We interpret the destination operator in a graphicable algebra

$$D_i = \sum_{k=1}^{\infty} \rho_i L (\rho_i^0 L)^{k-1}.$$

Suppose we start from x and our destination is vertex e_i. Now we want to study the number of paths that can be possibly taken:

- In one step, the number of paths from x to e_i is given by $\rho_i L(x)$.
- In two steps when the first arrival at e_i happens in the second step, the number of paths from x to e_i is given by $\rho_i L(\rho_i^0 L)(x)$.
- In three steps when the first arrival at e_i happens in the third step, the number of paths from x to e_i is given by $\rho_i L(\rho_i^0 L)^2(x)$.
- More generally, in n steps when the first arrival at e_i happens in nth step, the number of paths from x to e_i is given by $\rho_i L(\rho_i^0 L)^{n-1}(x)$.
- Within n steps, the total number of paths from x to e_i is given by

$$D_i^n = \sum_{k=1}^{n} \rho_i L (\rho_i^0 L)^{k-1}$$

Definition 18. *1) A geodesic from vertex x_i to vertex x_j is defined to be any path l with the minimum length between x_i and x_j,*

$$Length(l) = \min \left\{ k : \rho_{x_j} L^k(x_i) \neq 0 \right\}.$$

2) We call two generators x_i, x_j *adjacent in graphicable algebras, if* $\rho_{x_j} L(x_i) \neq 0$ *or* $\rho_{x_i} L(x_j) \neq 0$.

3) Cycle algebras: for a circle Fig. 6.1 with p vertices
we define a cycle algebra of dimension p to be an evolution algebra with generator set $\{x_1, x_2, \cdots, x_p\}$ and the defining relations are given

$$x_1^2 = x_p + x_2, \ x_2^2 = x_1 + x_3, \ \cdots,$$
$$x_{p-1}^2 = x_{p-2} + x_p, \ x_p^2 = x_1 + x_{p-1}.$$

4) Path algebras: for a path Fig. 6.2 with p vertices
we define a path algebra of dimension p to be an evolution algebra with generator set $\{x_1, x_2, \cdots, x_p\}$ and the defining relations are given

$$x_1^2 = x_2, \ x_2^2 = x_1 + x_3, \ x_3^2 = x_2 + x_4, \cdots$$
$$x_{p-2}^2 = x_{p-3} + x_{p-1}, \ x_{p-1}^2 = x_{p-2} + x_p, \ x_p^2 = x_{p-1}.$$

5) Complete algebras: for a complete graph Fig. 6.3 with p vertices
we define a complete algebra of dimension p to be an evolution algebra

$$x_i^2 = \sum_{k \neq i} x_k, \ i = 1, 2, \cdots, p.$$

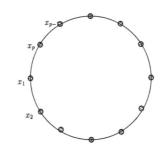

Fig. 6.1. Circle graph

Fig. 6.2. Path graph

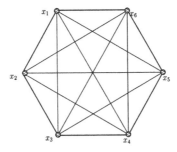

Fig. 6.3. Complete graph

Theorem 44. *(A classification of directed graphs) All directed graphs can be classified by their associated evolution algebras up to the skeleton-shape of evolution algebras.*

6.2 Further Research Topics

6.2.1 Evolution algebras and graph theory

For directed graphs, we have obtained a classification theorem by using our evolution algebras. The next main problem that has practical significance in here is, for a given directed graph, how to find the heads of the hierarchical structure algorithmically. Evolution algebras are very interesting and important because in evolution algebras, graphs can be represented in the form of algebras so that properties of graphs can be studied from the perspective of algebras. In this way, evolution algebras will become a systemized method to study graphs. This algebraic method has conspicuous advantages over the combinatorial method in graph theory. One question we should dig into first is whether every statement or problem in graph theory can be translated into the language of evolution algebras. If this is indeed the case, we will have a brand new "algebraic graph theory" and it will bring new and significant prospect in studying computer science. A well-known fact about combinatorics is that it lacks a systematized method to study despite its importance in application. Gian-Carlo Rota' hoped about combinatorics – "Combinatorics needs fewer theorems and more theory" [29]. Evolution algebra theory may direct combinatorics toward a trend that will be more systematized and more theory-like.

Using evolution algebras to study random graphs and networks is another interesting topic. As a matter of fact, random graphs and networks are the study of the processes of randomly forming graphs with given vertices. It is also important to study the evolution of random graphs and networks. By using evolution algebras, the study of the evolution of random graphs and networks can be transformed into the study of the evolution of evolution algebras. Obviously, the algebraic study of the evolution of random graphs and networks through using evolution algebras will systematically and fundamentally simplify the way of study.

Here we list some questions that are very interesting to study:

1. For all types of graphs, we can have their corresponding evolution algebras. The question is whether every statement, including even hard-problems in graph theory, can be translated to an algebraic statement.
2. For the first question, if the answer is "yes," then, can we solve some hard-problems by using evolution algebras?
3. As we know, nonassociative algebras are not easy to study. Graph theory can provide a tool to study them by using evolution algebras.

4. It is very interesting to study the relationship between random walks on a graph and the evolution algebra determined by the graph. This will add a new landscape in discrete geometry.

5. Random graphs and their evolution algebras: since theory of random graphs studies processes of forming graphs, given vertices, we may consider them as a type of evolution algebras, whose structure coefficients are random variables under a certain probability distribution. This kind of perspective will surely give insight into the study of networks.

6. When considering weighted graphs from the perspective of evolution algebras, all the above problems are also worthwhile and applicable to study.

7. There is a natural correspondence between evolution algebras and direct graphs. Direct graph theory is a good tool to study nonassociative algebras. The question is how good the tool is and how far we can go in this direction.

6.2.2 Evolution algebras and group theory, knot theory

Any group can be associated with an evolution algebra. Specifically, let G be a group, E be a finite set of generators of the group G, and K be a field. To define an evolution algebra, we take G as the generator set, and defining relations are:

$$g * g = \sum_{e \in E} k_e ge,$$

and

$$g * h = 0,$$

where $*$ is the algebraic multiplication, $k_e \in K$. One question is how the evolution algebra reflects properties of the corresponding group. For example, let G be a cyclic group with generator g, and we take the underlying field to be the real number field R. If we define

$$g^r * g^r = g^r(g + g^{-1}) = g^{r+1} + g^{r-1.},$$

after a computation, we have

$$(g^r)^{[n]} = \sum_{k=0}^{n} \binom{k}{n} g^{r+n-2k}.$$

Now if the group G is infinite, we can see, except the unit element of the group, each element of the group has a period 2. If the group has order m, except the unit, each element has a period $2 + m$. Therefore, infiniteness of cyclic group can be reflected by the period of nonunit elements in the evolution algebra. A very interesting case is for braid groups [27], which are the fundamental algebraic structures behind 3-manifolds and knot theory. We need to study whether anything deep about knots can be obtained from the associated evolution algebras.

6.2.3 Evolution algebras and Ihara-Selberg zeta function

For a finite graph X, we have the Ihara-Selberg zeta function $Z(u)$:

$$Z(u) = \prod_{w \in W} (1 - u^{|w|})^{-1},$$

where W denote the set of all prime walks in X, $|w|$ denote the steps of the walk w. When a graph is viewed as an evolution algebra, we will have an algebraic version of Riemann-zeta function. It is very interesting to study for what kind of evolution algebra Riemann hypothesis holds.

6.2.4 Continuous evolution algebras

Continuous versions of evolution algebras will be very interesting because they have a kind of semi-Lie group structure. One way to define continuous evolution algebras is to take structural coefficients to be differential functions over the underlying field. For example, if the generator set is $\{e_i | i \in \Lambda\}$, the defining relations are

$$e_i \cdot e_i = \sum_j a_{ij}(t) e_j,$$

where $a_{ij}(t)$ is a differential function; $e_i \cdot e_j = 0$. They will have relations with continuous-time Markov chains, just as discrete versions of evolution algebras have relations with discrete-time Markov chains. It is expected that continuous evolution algebras will be a powerful tool to study evolution of dynamical systems.

6.2.5 Algebraic statistical physics models and applications

Evolution algebras, either discrete or continuous, can be applied to study many real-world models. For example, in the study of "multi-person simple random walks" [26], we fix a graph G, and there are any finite number of persons distributed randomly at the vertices of G. We run a discrete time Markov chain with these persons over the graph. In each step of the Markov chain, we randomly pick up a person and move it to a random adjacent vertex. The problem we are concerned is the expected number of steps for these persons to meet all together at a specific vertex. Although we introduced tensor powers of a graph and a tensor products of a Markov chain to study this problem, we found evolution algebras are very helpful in making computations more clear. Therefore, we start to think that evolution algebras are useful in describing statistical physical problems, such as interactions, and absorption of particles. It may be necessary to study evolution algebra versions of statistical physics models, such as percolation processes, Ising models, Ashkin-Teller-Potts models, etc. We hope to achieve an algebraic theory of interaction of particles. Definitely, it will give us insight into various questions that are related to concepts in statistical physics, including computational complexity.

To achieve an algebraic theory of interaction of particles, we have started some work. A general multiplication is necessary for study of particle interactions. In this case we need to consider a multiplication of three-dimensional matrices. Suppose $A = (a_{ijk})$ be a three-dimensional matrix, and then we define $A \cdot A = (b_{ijk})$ to be a three-dimensional matrix whose entry is given by

$$b_{ijk} = \sum_{s,t} a_{stk} a_{ijt} a_{ijs}.$$

6.2.6 Evolution algebras and 3-manifolds

As mentioned in Chapter 2, for a triangulation t_1 of a 3-manifold M, we can define an evolution algebra $A(M, t_1)$. When taking the barycentric subdivision t_2 of t_1, we will have another evolution algebra $A(M, t_2)$. We can keep this procedure to get an infinite sequence of evolution algebras

$$\{A(M, t_n)\}_{n=1}^{\infty}.$$

Now, a lot of interesting questions arise: what is the limit of this sequence? what kind of properties of the 3-manifold can be reflected by this sequence of algebras? We conjecture that the limit of this sequence is closely related to the Laplace-Beltrami operator of 3-manifold M. Whatever it is, this is an open area that is interesting to explore.

6.2.7 Evolution algebras and phylogenetic trees, coalescent theory

Phylogenetic trees can be corresponded to a class of evolution algebras. This class of algebras has an algebraic regularity, which is the reflection of bifurcation of trees. If we could construct the evolution algebras from the data of a specific species or related many species, we can predict certain properties about the underlying evolution. Evolution algebras could also be used to study the genetic evolution reversely over time, namely, coalescent theory [23] [25]. The rationale of utility of evolution algebras in genetic evolution is that the mathematical objects of genetic evolution are discrete spaces, graphs, or graph-like spaces.

6.3 Background Literature

Our research goal is to use mathematics, including topology, algebra, differential equation, probability, stochastic processes, statistics, etc., to study and promote our understanding of natural phenomena, particularly biological and physical phenomena, and artificial complex systems. We also hope to look for new mathematical structures and new mathematical subjects inspired by scientific problems. In the research of the book, we have studied several fields, such as general genetic algebras, nonassociative algebras, and stochastic processes including Markov chains. Although the direct related literature about the subject that we are addressing in this book is not transparently seen anywhere, our background literature should still be referred to general genetic algebras and stochastic processes.

As to algebras in genetics, there are two comprehensive books that are worthwhile to be referred to. One is "Algebras in Genetics" written by Angelika Wörz-Busekros [46], the other is "Mathematical structures in population genetics" written by Lyubich, Y. I [47]. There are lots of reference papers in these two books. A good survey article is Reed's "Algebraic Structure of Genetic Inheritance" [48] published in the Bulletin of The American Mathematical Society. In our paper "Coalgebraic structure of genetic inheritance" [22], we trace back the idea to an Augustinian Monk Gregor Mendel, who began to discover the mathematical character of heredity. First, in his paper "Experiments in Plant-Hybridization" [30], Mendel employed some symbolism, which is quite algebraically suggestive, to express his laws. In fact, it was termed "Mendelian algebras" by several authors later. In the twenties and thirties of the last century, general genetic algebras were introduced. Apparently, Serebrowsky (in his paper "On the properties of the Mendelian equations" [31]) was the first to give an algebraic interpretation of the sign "×," which indicated sexual reproduction, and to give a mathematical formulation of Mendel's laws. Since then, the dynamics of reproduction of organisms was introduced to mathematics as a multiplication in algebras. The systematic study of algebras occurring in genetics was due to I.M.H. Etherington. In his series of seminal papers "Non-associative algebra and the symbolism of genetics" [34], he succeeded in giving a precise mathematical formulation of Mendel's laws in terms of nonassociative algebras. When we consider asexual inheritance, which is not Mendelian inheritance, the algebraic structure that we study in the present book appears. Besides the reference in genetic algebras, we also studied general biology from [11] [14] [15].

Stochastic processes including Markov chains have been very hot, and extensively studied fields and many theoretical and practical problem have been studied in this fields. Because of the limited time and space, we just list some of the literatures in the fields that we have studied and referred to in bibliography [1] [13] [20] [21]. However from our knowledge, there was no one to reveal the dynamical hierarchy of a general Markov chain, particularly the algebraic structure of a Markov chain in the literature. Research have generally been focused on a particularly irreducible Markov chain or on a particularly transient Markov chain, and then on computing some "interesting" quantity. This is not a good "pure mathematics" in the eyes of a "pure mathematician." Once we turn Markov chains into algebras, we can see a whole picture about the theory of Markov chains. We can also have a new classification for Markov chains. Because Markov chains model a lot of natural phenomena, our evolution algebra will add certain deep understanding about the structures of these natural phenomena.

Because the idea of taking the multiplication in algebra as a dynamical step comes from our study of general genetic algebras and stochastic processes, general references, besides the references about general genetic algebras and Markov chains, is about theory of nonassociative algebras. The book "An introduction to nonassociative algebras" [3] written by Richard Schafer is a good introductory book. We have also listed some books about nonassociative algebras, which we read and used in our research [4] [6] [7] [8] [9] [10] [19].

We also took some ideas from graph theory by reading some related books in this subject [5] [12] [18]. So we also list them in bibliography. I hope I can credit every author who contributes to my research in this book.

References

1. Dean Isaacson & Richard Madsen, Markov Chains Theory and Applications, John Wiley & Sons, New York, 1976.
2. Warren Ewens, Mathematical Population Genetics, Springer-Verlag, Berlin, 1979.
3. Richard Schafer, An Introduction to Nonassociative Algebras, Dover Publications, Inc. New York, 1994.
4. Roberto Costa, Alexander Griohkov, Henrique Quzzo, Luiz Peresi (edited), Nonassociative Algebra and its Applications, the Fourth International Conference, Marcel Dekker, Inc. 2000.
5. Chris Godsil, Gordon Royle, Algebraic Graph Theory, Springer, 2001.
6. Jack Lohmus, Eugene Paal, Leo Sorgsepp, Nonassociative Algebras in Physics, Hadronic Press, Inc. 1994.
7. Santas Gonzalez (edited), Non-Associative Algebra and Its Applications, Kluwer Academic Publishers, Dordrecht, 1994.
8. Kevin McCrimmon, A Taste of Jordan Algebras, Springer, New York, 2004.
9. A. I. Kostrikin, I. R. Shafarevich, Algebra VI, Springer, Berlin, 1995.
10. Susumu Okubo, Introduction to Octonion and Other Non-Associative Algebras in Physics, Cambridge University Press, 1995.
11. John Gillespie, Population Genetics, a Concise Guide, the Johns Hopkins University Press, Baltimore, 1998.
12. Marshall Hall, Combinatorial Theory, John Wiley & Sons, New York, 1986.
13. Mark Freidlin, Markov Processes and Differential Equations: Asymptotic Problems, Birkhauser Verlag, Basel, 1996.
14. Francisco Ayala, John Kiger, Modern Genetics, the Benjamin Cummings Publishing Company, Inc. 1980.
15. Nell Campbell, Biology, fourth edition, the Benjamin Cummings Company, Inc. 1996.
16. Scott Freeman & Jon Herron, Evolutionary Analysis, Prentice Hall, 1998.
17. Noboru Nakanishi, Graph Theory and Feynman Integrals, Gordon and Breach Science Publishers, New York, 1971.
18. Reinhard Diestel, Graph Theory, second edition, Springer, New York, 2000.
19. Nathan Jacobson, Structure and Representations of Jordan Algebras, AMS Colloquium Publications, 1982.

20. John G. Kemeny, J. Laurie Snell, Anthony W. Knapp, Denumerable Markov Chains, Springer-Verlag, New York, 1976.
21. Kai Lai Chung, Lectures from Markov Processes to Brownian Motion, Springer-Verlag, New York, 1982.
22. Jianjun Tian & Bai-Lian Li, *Coalgebraic structure of genetic inheritance*, Mathematical Bioscience and Engineering, vol.1, **2**, pp243-266, 2004.
23. Jianjun Tian & Xiao-Song Lin, *Colored coalescent theory*, Discrete and Continuous Dynamical Systems, pp833-845, 2005.
24. Jianjun (Paul) Tian & Petr Vojtechovsky, *Mathematical concepts of evolution algebras in non-Mendelian genetics*, Quasigroup and Related System, Vol.24, pp111-122, 2006.
25. Jianjun Tian & Xiao-Song Lin, *Colored genealogical trees and coalescent theory*, submitted.
26. Jianjun (Paul) Tian & Xiao Song Lin, *Continuous-time Markov process on graphs*, Stochastic Analysis and Applications, Vol.24, **5**, pp953-972, 2006.
27. Jones, V. *Hecke algebra representations of braid groups and link polynomials*, Ann. Math. 126(1987), 335-388.
28. Lin, X.-S. Tian, F. & Wang, Z. *Burau representation and random walks on string links*, Pacific J. Math. 182. **2**, 289-301, 1998.
29. Graver, J. E. & Watkins, M. E. Combinatorics with Emphasis on the Theory of Graphs, Springer-Verlag, New York, 1977.
30. Mendel, G., *Experiments in plant-hybridization*, Classic Papers in Genetics, pages 1-20, J. A. Peter editor, Prentice-Hall Inc. 1959.
31. Serebrowsky, A., *On the properties of the Mendelian equations*, Doklady A.N.SSSR. **2**, 33-36, 1934 (in Russian).
32. Glivenkov, V., *Algebra Mendelienne comptes rendus* (Doklady) de l'Acad. des Sci. de l'URSS **4**, (13), 385-386, 1936 (in Russian).
33. Kostitzin, V.A., *Sur les coefficients mendeliens d'heredite*, Comptes rendus de l'Acad. des Sci. **206**, 883-885, 1938 (in French).
34. Etherington, I.M.H., *Non-associative algebra and the symbolism of genetics*, Proc. Roy. Soc. Edinburgh B **61**, 24-42, 1941.
35. Gonshor, H., *Contributions to genetic algebras*, Proc. Edinburgh Math. Soc (2), 273-279, 1973.
36. Schafer, R.D., *An introduction to non-associative algebras*, Acad. Press, New York, 1966.
37. Holgate, P., *Sequences of powers in genetic algebras*, J. London Math., **42**, 489-496, 1967.
38. Holgate, P., *Selfing in gentic algebras*, J. Math. Biology, **6**, 197-206, 1978.
39. Hench, I., *Sequences in genetic algebras for overlapping generations*, Proc. Edinburgh Math. Soc. (2) **18**, 19-29, 1972.
40. Reiser, O., *Genetic algebras studied recursively and by means of differential operators*, Math. Scand. **10**, 25-44, 1962.
41. Abraham, V.M., *Linearising quadratic transformations in genetic algebras*, Thesis, Univ. of London, 1976.
42. Kempthorne, O., An Introduction to Genetic Statistics, John Wiley & Son, Inc., New York, 1950.
43. Sloane, N.J.A., and Wyner, A.D., (editors) Claude Elwood Shannon Collected Papers, IEEE Press, Piscataway, NJ. 1993.
44. Cotterma, C.W., A calculus for statistico-genetics, Dissertation, The Ohio State University, Columbus, OH. 1940.

45. Ballonoff, P., Genetics and Social Structure, Dowden, Hutchinson & Ross, Stroudsburg, PA. 1974.

46. Worz-Busekros, A., Algebras in Genetics, Lecture Notes in Biomath. 36, Springer-Verlag, Berlin, 1980.

47. Lyubich, Y.I., Mathematical Structures in Population Genetics, Springer-Verlag, New York, 1992.

48. Reed, M.L., *Algebraic structure of genetic inheritance*, Bull. of AMS, 34, (2), 107-130, 1997.

49. Baur, E., Zeit. Vererbungsl. **1**, 330-351, 1909.

50. Correns, C., Zeit. Vererbungsl. **1**, 291-329, 1909.

51. Birky, C.W.Jr., *The inheritance of genes in mitochondria and chloroplasts: laws, mechanisms, and models*, Annu. Rev. Genet. **35**, 125-148, 2001.

52. Birky, C.W.Jr., *Inheritance of mitochondrial mutations*, Mitochondrial DNA Mutations and Aging, Disease and Cancer, K.K. Singh, edit, Spring, 1998

53. Gillham, N.W., Organelle Genes and Genomes, Oxford University Press, 1994

54. Emmerson, C.F., Brown, G.K. and Poulton, J., *Synthesis of mitochondrial DNA in permeabilised human cultured cells*, Nucleic Acids Res. 29, **2**, 2001.

55. Ling F. and Shibata, T., *Mhr1p-dependent concatemeric mitochondrial DNA formation for generating yeast mitochondrial homoplasmic cells*, Mol. Biol. Cell, vol.15, 310-322, Jan. 2004.

56. Tang, Y., Manfredi, G., Hirano, M. and Schon, E.A., *Maintenance of human rearranged mitochondrial DNAs in long-term transmitochondrial cell lines*, Mol. Biol. Cell, vol.11, 2349-2358, Jul. 2000

57. Samen, F.M.A., Secor, G.A., and Gudmestad, N.C., *Variability in virulence among asexual progenies of Phytophthora infestans*, Phytopathology, **93**, 293-304, 2003.

58. Fry, W.E. and Goodwin, S.B., —textitRe-emergence of potato and tomato late blight in the United States, Plant Disease, 1349-1357, dec. 1997.

59. Goodwin, S.B., Cohen, B.A., and Fry, W.E., *Panglobal distribution of a single clonal lineage of the Irish potato famine fungus*, Proc. Natl. Acad. Sci. USA, **91**, 11591-11595, 1994.

60. Goodwin, S.B., *The population genetics of Phytophthora*, Phytopathology, **87**, 462-473, 1997.

61. Flor, H.H., *Current status of the gene for gene concept*, Annu. Rev. Phytopathol. **9**, 275-296, 1971.

Index

Lecture Notes in Mathematics

For information about earlier volumes
please contact your bookseller or Springer
LNM Online archive: springerlink.com

Vol. 1781: E. Bolthausen, E. Perkins, A. van der Vaart, Lectures on Probability Theory and Statistics. Ecole d' Eté de Probabilités de Saint-Flour XXIX-1999. Editor: P. Bernard (2002)

Vol. 1782: C.-H. Chu, A. T.-M. Lau, Harmonic Functions on Groups and Fourier Algebras (2002)

Vol. 1783: L. Grüne, Asymptotic Behavior of Dynamical and Control Systems under Perturbation and Discretization (2002)

Vol. 1784: L. H. Eliasson, S. B. Kuksin, S. Marmi, J.-C. Yoccoz, Dynamical Systems and Small Divisors. Cetraro, Italy 1998. Editors: S. Marmi, J.-C. Yoccoz (2002)

Vol. 1785: J. Arias de Reyna, Pointwise Convergence of Fourier Series (2002)

Vol. 1786: S. D. Cutkosky, Monomialization of Morphisms from 3-Folds to Surfaces (2002)

Vol. 1787: S. Caenepeel, G. Militaru, S. Zhu, Frobenius and Separable Functors for Generalized Module Categories and Nonlinear Equations (2002)

Vol. 1788: A. Vasil'ev, Moduli of Families of Curves for Conformal and Quasiconformal Mappings (2002)

Vol. 1789: Y. Sommerhäuser, Yetter-Drinfel'd Hopf algebras over groups of prime order (2002)

Vol. 1790: X. Zhan, Matrix Inequalities (2002)

Vol. 1791: M. Knebusch, D. Zhang, Manis Valuations and Prüfer Extensions I: A new Chapter in Commutative Algebra (2002)

Vol. 1792: D. D. Ang, R. Gorenflo, V. K. Le, D. D. Trong, Moment Theory and Some Inverse Problems in Potential Theory and Heat Conduction (2002)

Vol. 1793: J. Cortés Monforte, Geometric, Control and Numerical Aspects of Nonholonomic Systems (2002)

Vol. 1794: N. Pytheas Fogg, Substitution in Dynamics, Arithmetics and Combinatorics. Editors: V. Berthé, S. Ferenczi, C. Mauduit, A. Siegel (2002)

Vol. 1795: H. Li, Filtered-Graded Transfer in Using Noncommutative Gröbner Bases (2002)

Vol. 1796: J.M. Melenk, hp-Finite Element Methods for Singular Perturbations (2002)

Vol. 1797: B. Schmidt, Characters and Cyclotomic Fields in Finite Geometry (2002)

Vol. 1798: W.M. Oliva, Geometric Mechanics (2002)

Vol. 1799: H. Pajot, Analytic Capacity, Rectifiability, Menger Curvature and the Cauchy Integral (2002)

Vol. 1800: O. Gabber, L. Ramero, Almost Ring Theory (2003)

Vol. 1801: J. Azéma, M. Émery, M. Ledoux, M. Yor (Eds.), Séminaire de Probabilités XXXVI (2003)

Vol. 1802: V. Capasso, E. Merzbach, B. G. Ivanoff, M. Dozzi, R. Dalang, T. Mountford, Topics in Spatial Stochastic Processes. Martina Franca, Italy 2001. Editor: E. Merzbach (2003)

Vol. 1803: G. Dolzmann, Variational Methods for Crystalline Microstructure – Analysis and Computation (2003)

Vol. 1804: I. Cherednik, Ya. Markov, R. Howe, G. Lusztig, Iwahori-Hecke Algebras and their Representation Theory. Martina Franca, Italy 1999. Editors: V. Baldoni, D. Barbasch (2003)

Vol. 1805: F. Cao, Geometric Curve Evolution and Image Processing (2003)

Vol. 1806: H. Broer, I. Hoveijn. G. Lunther, G. Vegter, Bifurcations in Hamiltonian Systems. Computing Singularities by Gröbner Bases (2003)

Vol. 1807: V. D. Milman, G. Schechtman (Eds.), Geometric Aspects of Functional Analysis. Israel Seminar 2000-2002 (2003)

Vol. 1808: W. Schindler, Measures with Symmetry Properties (2003)

Vol. 1809: O. Steinbach, Stability Estimates for Hybrid Coupled Domain Decomposition Methods (2003)

Vol. 1810: J. Wengenroth, Derived Functors in Functional Analysis (2003)

Vol. 1811: J. Stevens, Deformations of Singularities (2003)

Vol. 1812: L. Ambrosio, K. Deckelnick, G. Dziuk, M. Mimura, V. A. Solonnikov, H. M. Soner, Mathematical Aspects of Evolving Interfaces. Madeira, Funchal, Portugal 2000. Editors: P. Colli, J. F. Rodrigues (2003)

Vol. 1813: L. Ambrosio, L. A. Caffarelli, Y. Brenier, G. Buttazzo, C. Villani, Optimal Transportation and its Applications. Martina Franca, Italy 2001. Editors: L. A. Caffarelli, S. Salsa (2003)

Vol. 1814: P. Bank, F. Baudoin, H. Föllmer, L.C.G. Rogers, M. Soner, N. Touzi, Paris-Princeton Lectures on Mathematical Finance 2002 (2003)

Vol. 1815: A. M. Vershik (Ed.), Asymptotic Combinatorics with Applications to Mathematical Physics. St. Petersburg, Russia 2001 (2003)

Vol. 1816: S. Albeverio, W. Schachermayer, M. Talagrand, Lectures on Probability Theory and Statistics. Ecole d'Eté de Probabilités de Saint-Flour XXX-2000. Editor: P. Bernard (2003)

Vol. 1817: E. Koelink, W. Van Assche (Eds.), Orthogonal Polynomials and Special Functions. Leuven 2002 (2003)

Vol. 1818: M. Bildhauer, Convex Variational Problems with Linear, nearly Linear and/or Anisotropic Growth Conditions (2003)

Vol. 1819: D. Masser, Yu. V. Nesterenko, H. P. Schlickewei, W. M. Schmidt, M. Waldschmidt, Diophantine Approximation. Cetraro, Italy 2000. Editors: F. Amoroso, U. Zannier (2003)

Vol. 1820: F. Hiai, H. Kosaki, Means of Hilbert Space Operators (2003)

Vol. 1821: S. Teufel, Adiabatic Perturbation Theory in Quantum Dynamics (2003)

Vol. 1822: S.-N. Chow, R. Conti, R. Johnson, J. Mallet-Paret, R. Nussbaum, Dynamical Systems. Cetraro, Italy 2000. Editors: J. W. Macki, P. Zecca (2003)

Vol. 1823: A. M. Anile, W. Allegretto, C. Ringhofer, Mathematical Problems in Semiconductor Physics. Cetraro, Italy 1998. Editor: A. M. Anile (2003)

Vol. 1824: J. A. Navarro González, J. B. Sancho de Salas, $\mathscr{C}^\infty$ – Differentiable Spaces (2003)

Vol. 1825: J. H. Bramble, A. Cohen, W. Dahmen, Multiscale Problems and Methods in Numerical Simulations, Martina Franca, Italy 2001. Editor: C. Canuto (2003)

Vol. 1826: K. Dohmen, Improved Bonferroni Inequalities via Abstract Tubes. Inequalities and Identities of Inclusion-Exclusion Type. VIII, 113 p, 2003.

Vol. 1827: K. M. Pilgrim, Combinations of Complex Dynamical Systems. IX, 118 p, 2003.

Vol. 1828: D. J. Green, Gröbner Bases and the Computation of Group Cohomology. XII, 138 p, 2003.

Vol. 1829: E. Altman, B. Gaujal, A. Hordijk, Discrete-Event Control of Stochastic Networks: Multimodularity and Regularity. XIV, 313 p, 2003.

Vol. 1830: M. I. Gil', Operator Functions and Localization of Spectra. XIV, 256 p, 2003.

Vol. 1831: A. Connes, J. Cuntz, E. Guentner, N. Higson, J. E. Kaminker, Noncommutative Geometry, Martina Franca, Italy 2002. Editors: S. Doplicher, L. Longo (2004)

Vol. 1832: J. Azéma, M. Émery, M. Ledoux, M. Yor (Eds.), Séminaire de Probabilités XXXVII (2003)

Recent Reprints and New Editions

4. Careful preparation of the manuscripts will help keep production time short besides ensuring satisfactory appearance of the finished book in print and online. After acceptance of the manuscript authors will be asked to prepare the final LaTeX source files (and also the corresponding dvi-, pdf- or zipped ps-file) together with the final printout made from these files. The LaTeX source files are essential for producing the full-text online version of the book (see http://www.springerlink.com/openurl.asp?genre=journal&issn=0075-8434 for the existing online volumes of LNM).

   The actual production of a Lecture Notes volume takes approximately 8 weeks.

5. Authors receive a total of 50 free copies of their volume, but no royalties. They are entitled to a discount of 33.3 % on the price of Springer books purchased for their personal use, if ordering directly from Springer.

6. Commitment to publish is made by letter of intent rather than by signing a formal contract. Springer-Verlag secures the copyright for each volume. Authors are free to reuse material contained in their LNM volumes in later publications: A brief written (or e-mail) request for formal permission is sufficient.

Addresses:

Professor J.-M. Morel, CMLA,
École Normale Supérieure de Cachan,
61 Avenue du Président Wilson, 94235 Cachan Cedex, France
E-mail: Jean-Michel.Morel@cmla.ens-cachan.fr

Professor F. Takens, Mathematisch Instituut,
Rijksuniversiteit Groningen, Postbus 800,
9700 AV Groningen, The Netherlands
E-mail: F.Takens@math.rug.nl

Professor B. Teissier, Institut Mathématique de Jussieu,
UMR 7586 du CNRS, Équipe "Géométrie et Dynamique",
175 rue du Chevaleret
75013 Paris, France
E-mail: teissier@math.jussieu.fr

For the "Mathematical Biosciences Subseries" of LNM:

Professor P. K. Maini, Center for Mathematical Biology,
Mathematical Institute, 24-29 St Giles,
Oxford OX1 3LP, UK
E-mail : maini@maths.ox.ac.uk

Springer, Mathematics Editorial, Tiergartenstr. 17,
69121 Heidelberg, Germany,
Tel.: +49 (6221) 487-8410
Fax: +49 (6221) 487-8355
E-mail: lnm@springer-sbm.com